Supported Reagents

Preparation, Analysis, and Applications

Supported Reagents

Preparation, Analysis, and Applications

James H. Clark

Adrian P. Kybett

Duncan J. Macquarrie

VCH

James H. Clark
Adrian P. Kybett
Duncan J. Macquarrie
Department of Chemistry
University of York
Heslington, York
YO1 5DD, U.K.

Library of Congress Cataloging-in-Publication Data

Clark, James H.
Supported reagents : preparation, analysis, and applications / James H. Clark, Adrian P. Kybett, Duncan J. Macquarrie.
p. cm.
Includes bibliographical references and index.
ISBN 1-56081-010-6
1. Supported reagents. I. Kybett, Adrian P. II. Macquarrie, Duncan J. III. Title.
QD77.C59 1992
543'.01—dc20 92-5775
CIP

British Library Cataloguing in Publication Data

Clark, James H.
Supported Reagents: Preparation and Analysis
I. Title
546

ISBN 1-56081-010-6

Printed in the United States of America.

ISBN 1-56081-010-6 VCH Publishers
ISBN 3-527-28043-X VCH Verlagsgesellschaft

Published jointly by:

VCH Publishers, Inc.
220 East 23rd Street
Suite 909
New York, New York 10010

VCH Verlagsgesellschaft mbH
P.O. Box 10 11 16
D-6940 Weinheim
Federal Republic of Germany

VCH Publishers (UK) Ltd.
8 Wellington Court
Cambridge CB1 1HZ
United Kingdom

We would like to express our thanks to
Liz Shields and May Price for all their help
in the preparation of this book.

Preface

Supported reagents have been known since Fetizon's original article describing silver carbonate-celite in 1968. The first comprehensive reviews on their applications in synthesis appeared in 1979 and the first entire book devoted to the subject was published in 1987. The last few years in particular have seen several significant breakthroughs in the subject including the discovery of truly catalytic supported reagents, their successful scale-up to an industrial scale, and the effective use of modern analytical techniques to probe the surfaces of the materials and to help optimize their performance in chemical reactions. The first International Symposium on Supported Reagent Chemistry was held in England in 1991 and it brought together chemists with many different backgrounds—in catalysis, organic synthesis, analysis and materials science, as well as from industry and academia. Added momentum to the development of the subject has come from new national and trans-national environmental legislation which is placing increasing demands on the chemical industry to replace outdated chemical processes with new cleaner technologies and thus reduce pollution at source. End-of-pipe solutions are now the least attractive option.

One important aspect of "Clean Technology" will be the use of "Environmentally Friendly Catalysis" typically involving the use of solid catalysts the use of which will lead to minimal pollution and waste and will themselves be environmentally benign. The application of such catalysis to fine chemicals manufacturing—an area where heterogeneous catalysis has traditionally been little used—is likely to be especially important in the future. Supported reagents should have an important role to play as environmentally friendly catalysts.

In many ways, supported reagent chemistry is coming of age and seems destined to play an increasingly important role in research and industrial process technology into the future. This book has been written as a guide to the chemist who is planning to use or is considering the possible use of supported reagents. One of the most difficult aspects of the subject is the number of choices available in the preparation; loading, support, method of preparation and activation are all important parameters. Chapters 1 and 2 focus the readers mind on the most important factors which need to be taken into account in selecting the particular combination of parameters most appropriate to the job involved. The importance of analysis in heterogeneous catalysis is well established but it is only in recent years that its value is improving the performance of supported reagents has become apparent. There are very many analytical techniques that can be applied to solids and Chapter 3 deals only with those which are most likely to be available and that give useful information quickly! As in Chapter 2, the emphasis is on the supported reagents rather than the methods or techniques and we are deliberately generous with our use of examples. Finally, in Chapter 4 three types of supported reagents which have received significant attention in the literature are used as case studies to help illustrate the various topics covered earlier.

This book should provide you with a unique guide and reference source for supported reagent chemistry. It should hopefully convince the reader to make use of these essentially simple but fascinating materials which in a little over twenty years are maturing from academic curiosities to important industrial materials.

J. H. C.
A. P. K.
D. J. M.

York, United Kingdom
April 1992

Contents

1 Introduction to Supported Reagent Chemistry

1.1. Definitions

The term *supported reagent* has been used to describe a wide range of materials including both inorganic and organic "supports." The most general definition of the term is the adsorption on, dispersion over, or intercalation in an insoluble support material of a chemical reagent.[1] The interaction between reagent and support may be either by chemical bonding (chemisorption) or by other physical modes (physisorption). This book is predominantly devoted to those supported reagents and catalysts prepared using inorganic and generally high surface area support materials. The most common of these are aluminas, clays, silicas, and zeolites. In general, the reagents considered are inorganic, although there are some notable organic exceptions to this. The supported reagents of interest are those that are of use in synthetic organic chemistry, particularly for the formation of fine organics. The use of heavy metals as support materials or as reagents are generally excluded from consideration here.

The subject of supported reagent chemistry is inherently difficult to search in the chemical literature or abstracts. There are several reasons for this, not the least being the diverse interests of the workers in the field covering the broad areas of theoretical, physical, inorganic, organic, and analytical chemistries. The use of the word *support* in titles or abstracts is not general or satisfactory as a search term. Search terms, such as *alumina* or *silica,* are impracticable, producing unwieldy results. Nevertheless, computer literature searching facilities may be useful when several search terms are combined.

A further reason for the difficulties involved in systematically searching the literature is that the supported reagent may not be of primary importance in the article. A particularly helpful starting point remains the review "Organic Synthesis Using Supported Reagents",[1] although the volume of material published since then is vast. A table of examples of the use of supported reagents is given in Appendix 1.

1.2. The Importance of Supported Reagents

There are several important factors that have influenced the great surge of work with supported reagents in organic synthesis. One major, and only recently appreciated, advantage of supported reagents is their potential as environmentally friendly alternatives to corrosive, wasteful, and difficult to dispose of conventional reagents.[64] Thus, clay-supported reagents have been used to replace aluminium trichloride for Friedel–Crafts catalysis.[2]

A number of general advantages of using supported reagents have been put forward. A supported reagent often has a high surface area and a layered or porous nature. These factors may have advantages to the reaction energetics, enabling reactions to be carried out rapidly and in high yields under mild conditions. In many cases the isomer selectivities of a reaction may be changed or enhanced by using a supported reagent. An example that demonstrates these effects is the nitration of estrone using clay-supported iron III nitrate (clayfen).[3]

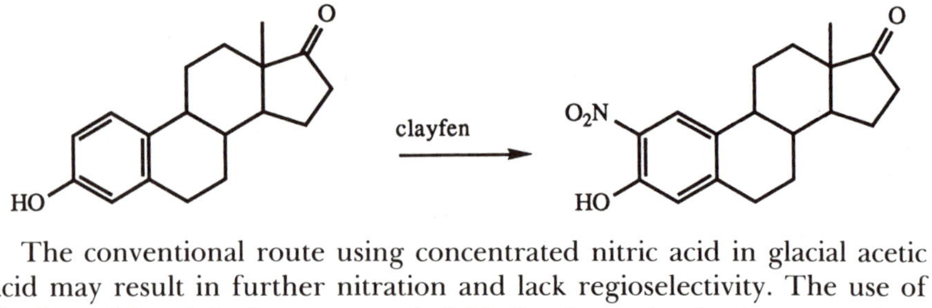

The conventional route using concentrated nitric acid in glacial acetic acid may result in further nitration and lack regioselectivity. The use of clayfen achieved the best reported yields for the regioselective mononitration illustrated.

The advantages of supported reagents in terms of reaction energetics may be accounted for in many ways. At the simplest level the benefit may be considered by the dramatic increase in surface area achieved by simply dispersing an inorganic reagent over a high surface area material. Thus, KF (which has a surface area of 1.3 m^2/g for spray dried samples)[4] when efficiently dispersed on alumina of surface area ca. 150 m^2/g

may have a greater than 100-fold increase in its accessible sites. This is a vital consideration, especially for reagents that are insoluble in organic media.

The general effect of reduced dimensionality in heterogeneous catalysis is also vital in supported reagent chemistry. Thus, substrate and reagent are more likely to meet as a result of random motion in a two-dimensional space rather than in three dimensions. The consideration of fractal dimension *(D)* of a support material may lead to a qualitative assessment of its effectiveness, solids of lower *D* constraining diffusion of reactants more effectively. This principle has been described by Laszlo,[5] who also quotes a range of *D* values from 2.9 for silica gel (porous solid) to 1.9 for activated charcoal (layered solid).[5]

Furthermore, the surface area and the fractal dimension must be considered in combination with the actual chemical nature of support surface. The most important features of concern here are usually the Brönsted and Lewis acid sites, as well as the degree of hydration of the surface.[64] For example, alumina normally contains a large amount of physisorbed water, also polarized surface =A1-O-H groups providing Brönsted acidity, as well as the expected Lewis acid sites of the alumina.

A charcoal surface, although having a lower fractal dimension and a higher surface area, may in some cases be less effective as a support material because of its lack of OH sites. Furthermore, the activity of any supported reagent is determined by its preparation (discussed in detail in Chapter 2). It may not be possible in practice to achieve actual monolayer dispersion of a reagent over a surface to take full advantage of the benefits described earlier. Scanning electron microscopy has shown that alkali metal fluorides form microclusters on alumina surfaces, but, although an increased effective surface area is achieved, it is not actually as high as the support material itself[6] (see Chapter 3). In addition, internal microporous surfaces may be inaccessible to reagents owing to steric constraints.

It may be reasonably argued that supported reagent chemistry has its origins in pure serendipity. For example, water present in chromatographic silica gel may function as an inefficient supported reagent, causing reactions where simple separations were intended. The development of the subject is such that careful design and characterization may achive highly efficient and selective catalysis. An example of this latter case is a new solid acid catalyst prepared by reacting an inorganic oxide with aluminium tri-chloride in refluxing carbon tetrachloride.[7] In this case the careful choice of a solvent for preparation resulted in a catalyst with considerably greater activity than that produced using other solvents (Table 1.1). In this case infrared and NMR spectroscopies demonstrated the active species to be surface (-O-)$AlCl_2$. This species is stable and does not evolve $AlCl_3$ from the surface.

A further example of the formation of a highly acidic supported reagent is the use of molybdenum oxide supported on zirconia.[8] This solid super-

Table 1.1 Relative Activities of $AlCl_2$-Functionalized Supports in the Cracking of Hexadecane

Support	Solvent	Relative Activity
Al_2O_3	CCl_4	5.9×10^6
Zeolite[a]	CCl_4	1.3×10^7
B_2O_3	CCl_4	6.1×10^6
TiO_2	CCl_4	5.0×10^4
MgO	CCl_4	7.8×10^3
Al_2Cl_6[b]	CCl_4	1.1×10^7
SiO_2	CCl_4	7.5×10^6
SiO_2	$CHCl_3$	8.2×10^3
SiO_2	CH_2Cl_2	1.1×10^2
SiO_2	$ClCH_2CHCl$	2.4
SiO_2	C_6H_6	1.0

[a] High silica zeolite.
[b] Unsupported Al_2Cl_6.

acid was prepared by impregnating H_2MoO_4 in aqueous ammonia onto $Zr(OH)_4$. Drying and air calcining the resulting solid produced an active catalyst for the liquid phase acylation of toluene with benzoic anhydride. The MoO_3–ZrO_2 catalyst was characterized using X-ray diffraction (XRD) and compared with a preparation commencing with the zirconium oxide. The XRD patterns were significantly different and zirconium oxide was inactive in the benzoylation. The authors also reported that the use of TiO_2, SiO_2, Al_2O_3, and MgO as support materials for the molybdenum oxide result in inactive reagents in the Friedel–Crafts reaction. These observations illustrate the importance of the choice of the support, the preparation method, and the characterization of the material, as well as the great potential of supported reagents. The acylation described generally requires strong acid catalysis, creating handling and disposal problems.

An interesting example of the selectivity achievable using support materials is the preferential esterification of longer chain dicarboxylic acids in the presence of shorter chain molecules.[9] In this case there is an interesting reversal of the usual supported reagent concept. The selectivity is in fact achieved by supporting the reactant while the reagent (diazomethane) remains in solution. One end of the diacid absorbs onto the alumina surface, leaving the opposite end of the molecule free to esterify. The greatest selectivity was found for mixtures of two diacids with the greatest difference in chain length. For example, the authors report that a mixture of C_{12} and C_5 diacids gave 89% formation of the monomethyl C_{12} ester with only 9% of the monomethyl C_5 ester.

A similar concept to the one described earlier has been applied to the methylation of alcohols and phenols with diazomethane.[10] Alcohols and phenols are adsorbed onto silica gel either by a strong hydrogen bond or by formation of the alkoxide. Here, the silica gel replaces the more conventional Lewis acid catalysts in the quantitative formation of methyl ethers using (in this case) gaseous diazomethane. Again, there is a strong dependence on the choice of support with alumina and titania promoting the reaction to a lesser extent. Zeolite, celite, and zinc oxide were less effective and magnesium oxide was totally inactive. The esterification and methylation reactions described are important reminders that the support materials themselves should not be regarded as being simply inert when actual supported reagents are considered.

The formation of supported ionenes on silica gel illustrates another useful attribute of supported reagents. Ionenes (polycations where quaternary ammonium functions form part of a polymer chain, or side groups on a chain) are generally hydrophilic and are used as catalysts in aqueous solutions. It has been reported, however, that supporting ionenes on silica gel enables them to be used as phase transfer catalysts in toluene solutions.[11] This enabled a number of substitution reactions, such as:

$C_{10}H_{21}OMs$ + ionene, Br^-/silica gel $\longrightarrow$
$C_{10}H_{21}Br$ + ionenes, , OMs^-/silica gel

There is some ambiguity in the formal definition of a supported reagent, but one reasonable category of materials to include is those that are formally ion exchanged. For example, considerable literature is available describing the use of ion-exchanged clays in organic synthesis. Thus, a methyl-*t*-butyl ether has been produced using either ion-exchanged resins or clays[12] and Friedel–Crafts alkylations have been accomplished using montmorillonite exchanged with transition metal cations.[13]

Zeolites have been used mostly as catalysts for large-scale, typically petrochemical syntheses; however, examples of applications in fine chemistry are available.[14] Zeolites are designated as support reagents with their acidic and basic properties as well as by virtue of supporting further reagents or catalysts. The acid strength of a zeolite may be altered by ion exchange and thermal treatments, and in a series of reactions it was possible to vary the products of the *t*-butylation of phenol according to zeolite acid strength.[15] This is a result of the *C*-alkylation requiring greater acidity than *O*-alkylation. All the acid sites are active for ether formation, whereas only a fraction of the Brönsted sites are active in *C*-alkylation. This fraction was found to increase exponentially with the extent of Na^+ exchange. The zeolite Si : Al ratio is also important in the determination of reaction products by zeolite catalysis. An example of this effect is described where HY zeolites (Na^+) only catalyze the alkylation of benzene by formaldehyde when the $pK_a > 0$.[16] In this case the authors describe the necessity for 0 or 1 Al to be in the next nearest site to the acid center for the reaction to proceed. This is an interesting example of extremely well defined supported reagents.

An example of the use of a reagent supported onto a zeolite is the use of *t*-butyl hypochlorite on zeolite X.[17] Here paraselective monochlorination of the aromatic compound was achieved under mild conditions.

Although this book concentrates on inorganic supports, it is important to recognize the wide use of organic materials as supports. For example, heterogeneous phase transfer catalysts, such as phosphonium salts or crown ethers, have been supported on polystyrene using long spacer chains.[18] In this way good catalytic activity was maintained while immobilizing them. Also, dimethyl sulfoxide has been supported on cross-linked polystyrenes as a phase transfer catalyst for the reaction of bromooctane with SCN^-, I^-, and CN^- as nucleophiles.[19] The cyanation of benzoyl chloride has been described where the cyanide is impregnated onto Amberlite resins.[20] This supported reagent was useful for the formation of acyl cyanides, but not aliphatic acyl cyanides. Indeed, it produced considerably superior reactivity and selectivity compared with alumina, silica, or activated charcoal as support materials.

The examples set out previously intended to show the variety of supported reagents described in the literature. Other ways of using supports and supported reagents exist, such as the in situ mixing of support (silica gel), reagent ($NaBH_4$), and reactant for the reduction of nitrostyrenes.[21] Also the existence of physisorbed non catalytic as well as chemisorbed catalytic sites in one supported reagent (dichromate or permanganate on alumina) has been described for the oxidation of diphenylmethane and other substrates.[22] Organometallic reactions have also been described that take place on the surface of a solid support:[23]

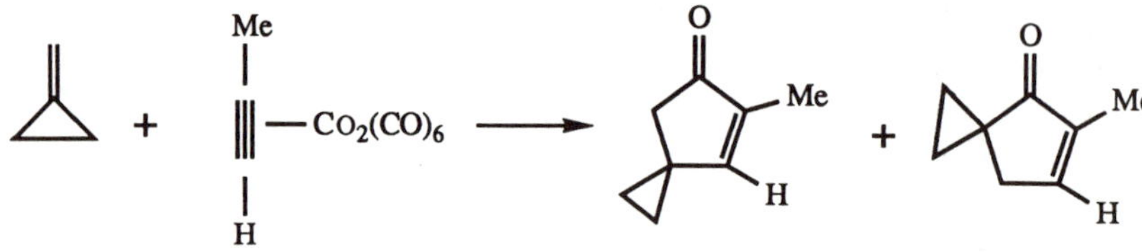

It has not been possible to absolutely define the term supported reagent. The chemistry considered has been wide ranging with a variety of supports, many reagents, and numerous categories of reactions promoted. Supported reagents have been shown to frequently enable increased yields and selectivities in synthesis.

1.3. Choice of Support Material

There are relatively few systematic explanations available to account for the successful combination of a support and a reagent, although numerous works describe comparisons between various supports for a particular reaction. A number of supports have been described in the previous section and in his 1979 review McKillop included celite, silica, alumina, graphite, carbon, montmorillonite K10 clay, Girdler KSF clay, molecular sieves (zeolites), and kieselguhr. Posner[24] reviewed the use of alumina reagents for organic reactions, and more recently, Cornelis and Laszlo[3] reviewed their own extensive work using montmorillonite reagents, and zeolite reagents have been reviewed by Perot and Guisnet.[14] The first supported reagent book also describes a variety of support materials and their uses.[25] Bram and co-workers[26] have also reviewed the use of a number of support materials for reduction.

The effectiveness of a particular support material for a given reagent and reaction is likely to be the result of a combination of factors including surface area, porosity, acidity–basicity, as well as the crystalline or amorphous nature of the material.[27] The surfaces of the materials under consideration here are generally irregular with voids, steps, pores, and other surface imperfections. In contrast, zeolites have well-defined crystalline characteristics. Particle size and shape contribute to the surface area of crystals or powders as does the material's porosity (usually defined as surface imperfections that are deeper than they are wide).

The physical adsorption of gases is the usual method of surface area determinations, although this may only measure external and not internal surfaces, as in the case of layered clays. Table 1.2 shows a variety of support surface areas. In each case a range of values is given, reflecting the great differences that may exist between samples.

Support pore structure is, in some cases (notably with zeolites), a critical factor in the choice of the support material. Specific channel structures may result in molecular shape selectivity where the size and the shape of a diffusing substrate or product are closely matched to the support pore diameter. The regioselective chlorination of toluene using a zeolite-supported reagent is an example of this effect.[17] Selectivity is also induced by the void structure of zeolites in decane hydroconversion.[32] In the case of silicas pore sizes range from less than 2 nm to greater than 200 nm.[29]

Clay (montmorillonites) and graphite are support materials with layered structures. With montmorillonite-supported reagents reactions may occur in the interlamellar region of the crystals. This situation is important in terms of the reduced dimensionality of the reaction (fractal dimen-

Table 1.2 Support Surface Areas

Support	Surface Area (m^2/g)	Reference
Alumina	120–300	28,30
Silica	50–420	29
Montmorillonite	10–270	30
Montmorillonite K10	220–270	30
Activated carbon	610–790	31

sion, described earlier) compared with a homogeneous reaction or one occurring on an external supported reagent surface. Montmorillonite may be pillared (layers permanently fixed apart) and thus achieve some degree of zeolitic character. For example, a large and sharp pore size distribution of ca. 3 nm was achieved by using an aluminium chlorohydroxy oligomer pillar in the presence of polyvinylalcohol (PVA). The PVA was used to wrap the oligomer to achieve a large layer swelling of the clay and was then removed by calcination.[33]

The careful choice of a support material is frequently critical to achieve satisfactory results in organic syntheses. It is also well established that the support chosen within a category may be important. For example, neutral, acidic, and basic aluminas from a variety of suppliers may give differing results.

An example of a systematic study of 10 support materials for sodium metaperiodate has been described.[34] Here, the identically prepared supported reagents were used for selective oxidations of sulfides to sulfoxides. The inorganic supports tested included alumina, celite, charcoal, florisil, montmorillonite clay, and silica gel, from which Merck acidic alumina was chosen for a series of sulfide oxidations.

In the preparation and subsequent oxidation of 2-nitroalcohols two different support materials were found effective to use for the two reactions.[35] Thus, KF supported on alumina was used for the condensation and CrO_3 on montmorillonite for the oxidation.

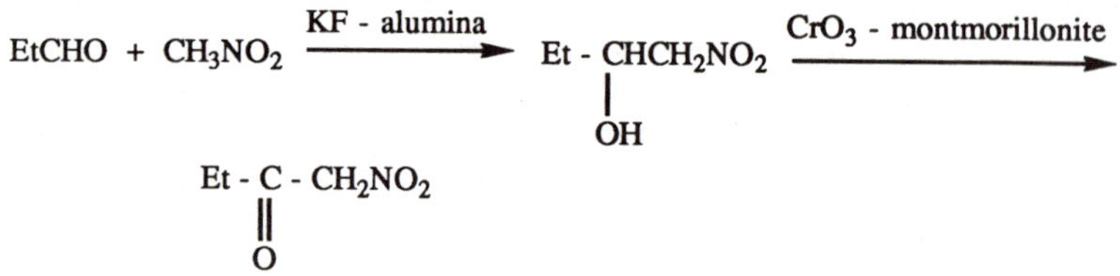

The KF–alumina reagent was chosen as an already established catalyst of aldol and Michael reactions. The use of alumina for the condensation reaction gave improved yields compared with homogeneous catalysts and the supported reagent was again superior to the untreated support. For the oxidation stage, montmorillonite was the support of choice as silica resulted in an unstable CrO_3 catalyst and CrO_3–alumina gave poor yields.

As mentioned in the previous example the support material can have catalytic activity itself. Alumina catalyzes the formation of iodooctane from bromooctane in the presence of NaI (also chlorination, acetylation, and cyanation).[36] In this case the use of silica gel may also promote the reaction, but with lower yields than alumina, 34% versus 95% for iodooctane formation. Elimination reactions (cyclohexyltosylates to cyclohexenes)[37] have been effected by dehydrated chromatographic alumina. Such a heterogeneous reagent was found to be suitable to introduce double bonds into polyfunctional compounds.[38]

The inorganic materials used as supports for catalyst preparations are often those used for chromatography, as such, alumina was found to promote an oxidation and reduction reaction on an eluted molecule (a Cannizzaro type reaction).[39]

$$2\ RCH_2CHO \xrightarrow{\text{Alumina}} RCH_2CH_2OH + (RCH_2CO_2)_3Al$$

R = [9-(*p*-methoxyphenyl)-9-fluorenyl]

It is noteworthy that this reaction does not proceed under normal base-catalyzed Cannizzaro reaction conditions.

Furthermore base-catalyzed condensation reactions have been promoted by neutral alumina (in preference to silica), such as the Knoevenagel reaction.[40]

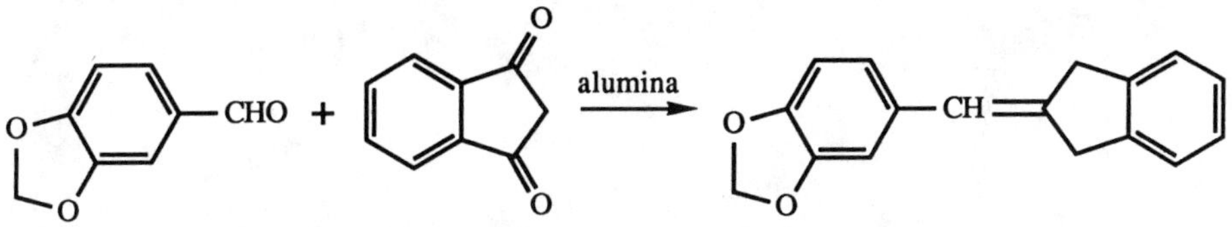

With less acidic compounds, KF–alumina was used.

A photochemical cleavage reaction has also been promoted by the inorganic support materials silica, $AlPO_4$, BPO_4–$AlPO_4$, and barium hydroxides.[41]

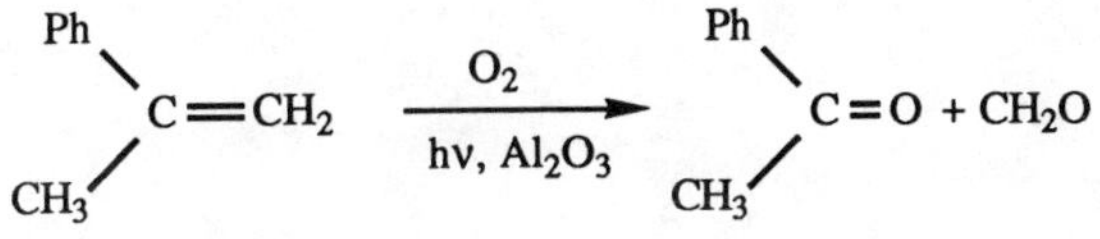

Here the microcrystalline BPO_4–$AlPO_4$ solids (of various compositions) were found to be more active than the amorphous silica or $AlPO_4$ materials. The reaction proceeded on either acidic or basic solids using electron acceptor and electron donor sites, respectively.

Table 1.3 describes a selection of reactions promoted by supported reagents and illustrates the relative effectiveness of particular support materials.

Not only can silica and alumina produce different types of reagent, variations in supported reagent can also be found using different versions of the same support material. For example, alumina is available in acidic, neutral, and basic varieties. The differences between these are due to the relative numbers of sites on the alumina surface. The surface of any alumina has amphoteric AlOH sites that can be converted into either basic or acidic sites (Fig.1.1). All aluminas contain some of each type of site and the differences between acidic, neutral, and basic aluminas stem from the amounts of each type of site, rather than from the presence or absence of

Table 1.3

Support	Reagent	Reaction	Relative Effectiveness	Comment	Reference
Alumina (acidic)	Na_2S	$PhCHO \rightarrow PhCH_2OH$	100	Acidic alumina more effective as measured by yield of alcohol	42
Alumina (neutral)			81		
Zeolite (H^+, Na^+, X)	^tBuOCl	$CH_3Ph \rightarrow$ 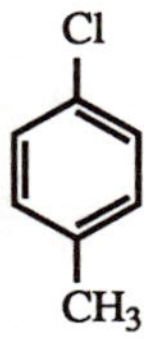	100	*para*-selectivity achieved using zeolite	17
Silica			43		
Alumina (neutral)	$K_2Cr_2O_7$	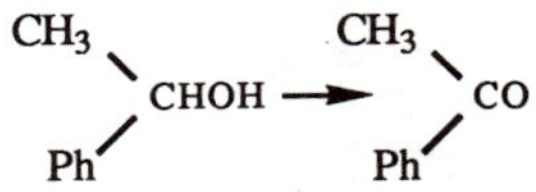	100	$K_2Cr_2O_7$–alumina is reagent of choice for various oxidations	43
Florisil			85		
Silica gel			79		
Alumina (neutral)	NaCN	$CH_3(CH_2)_6CH_2Br \rightarrow$ $\rightarrow CH_3(CH_2)_7CN$	100	15 different supports tested. Comparison of reactive cyanide	44
Molecular sieve (13X)			78		
Montmorillonite (K306)			22		
Molecular sieve (4A)			11		
Celite			6		
Molecular sieve (3A)	$KMnO_4$	Cyclododecanol → → Cyclododecanone	35	11 supports tested. Comparison of oxidizing capacity, mmole alcohol per gram reagent.	45
Molecular sieve (4A)			96		
Molecular sieve (5A)			43		
Molecular sieve (13X)			100		

Alumina (JRC-ALO-5) Alumina (Merck) Silica/Alumina (29% Al_2O_3)	Me_2SO_4	$Ph_2CHOH \rightarrow Ph_2CHOMe$	100 53 41		46
Silica Alumina Molecular sieve (5A) Molecular sieve (13X) Kieselgur	KSCN (10 mmol/g)	$PhCH_2Cl \rightarrow$ $PhCH_2\ SCN$	100 69 14 7 4		47
Alumina Silical Gel Molecular sieve (5A) Graphite	$CuCl_2$	PhOH → 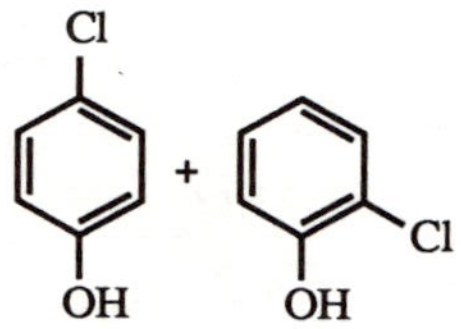	100 25 47 16	The alumina reagent gives best *p/o,* but it also gives a small amount of dichlorophenol.	48
Silica Alumina (γ) Carbon	$FeCl_3$	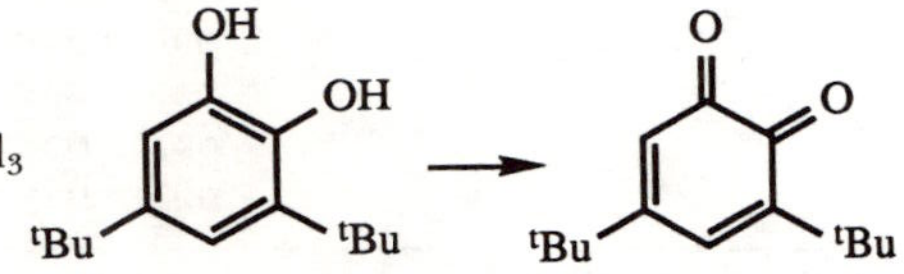	100 78 50		49
Silica Alumina	KSCN (2 mmol/g)	$PhCH_2Br \rightarrow PhCH_2SCN$	81 100	This low loading reverses the support effectiveness compared with the next example.	50
Silica Alumina	KSCN (10 mmol/g)	$PhCH_2Br \rightarrow PhCH_2SCN$	100 71	In close agreement with Ref. 47.	50
Silica CaF_2 Alumina	KSCN + I_2	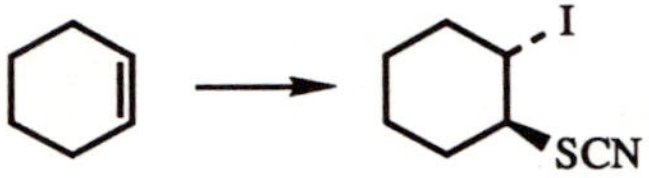	85 100 90		51

Table 1.3

Support	Reagent	Reaction	Relative Effectiveness	Comment	Reference
Al^{3+} Mineral colloid	ion exchanged	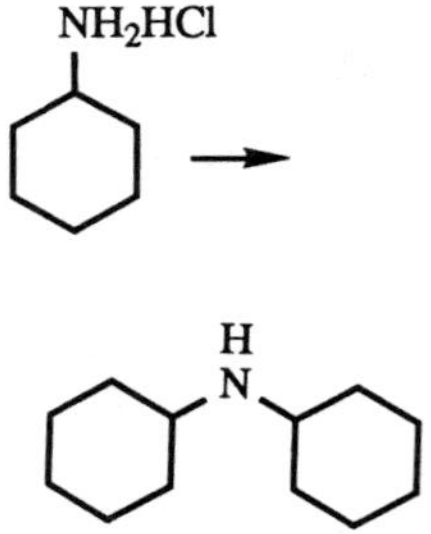	100	The relative reactivities of these sheet silicate catalysts is dependent on both the source of the clay and its ion exchange.	52
Al^{3+} Wyoming Bentonite	ion exchanged		91		
H^{+} Wyoming Bentonite	ion exchanged		87		
H^{+} Mineral colloid	ion exchanged		82		
Co^{2+} Wyoming Bentonite	ion exchanged		79		
Ni^{2+} Wyoming Bentonite	ion exchanged		78		
Mg^{2+} Wyoming Bentonite	ion exchanged		73		
Cr^{3+} Wyoming Bentonite	ion exchanged		72		
Al^{3+} Gelwhite	ion exchanged		69		
Al^{3+} Fullers Earth	ion exchanged		55		
Ca^{2+} Wyoming Bentonite	ion exchanged		35		
Gelwhite	non exchanged		30		
H^{+} Gelwhite	ion exchanged		24		
Na^{+} Wyoming Bentonite	non exchanged		22		
Mineral colloid	non exchanged		19		

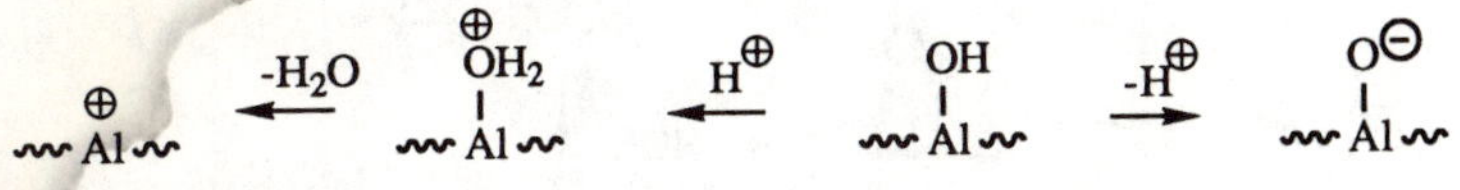

Figure 1.1 Surface sites of alumina.

one or more type of site. These differences will almost certainly affect the amount of chemisorption of species onto the surface as such processes are normally specific to one type of site. This should, therefore, affect the reactivity of the reagent. Indeed Clark et al.[53] found differences in the reactivity of dichromate–alumina-supported reagents prepared from different types of alumina. The catalytic activity of these materials was due to the chemisorbed dichromate, and variations in the nature of the alumina gave different levels of activity (Fig. 1.2). The compatibility of a surface with the reagent to be supported is an important factor in the preparation of active-supported reagents. Thus, the surface chemistry of the support must not cause decomposition of the molecules to be supported. For example, sodium cyanide reacts with acids to produce HCN, which is volatile and can be lost from the reaction system. This may partly explain the activity of a series of supported sodium cyanide reagents prepared by Regen et al.[2] They found that the most active reagents (i.e., those that contained the largest amount of reactive cyanide ion) were prepared from the relatively nonacidic neutral alumina and LMS–13X molecular sieves. Reagents prepared from acidic supports, such as K10 montmorillonite and silica, had only traces of reactive cyanide present. Other nonacidic support materials (e.g., LMS 3Å, 4Å, 5Å sieves) failed to give active reagents. This may be due to other factors, such as porosity or formation of clusters on external faces only, as is the case with $LiBH_4$ sieves.[3] Thus, it appears that surface acidity has a detrimental effect on the amount of reactive cyanide present in supported reagents owing to reactions such as (Fig. 1.3). The effect of support acidity is also exemplified by the work of Onaka et al.[55] They measured the surface acidity of three zeolites from which they prepared sodium azide–zeolite materials. Then they looked at the effect of surface acidity on the ring opening of epoxides using the

Figure 1.2 Effect of varying support on the efficiency of a supported reagent.

$$PhCH_2Ph \xrightarrow[O_2]{Na_2Cr_2O_7/Al_2O_3} PhCOPh$$

Support	Relative rate of oxidation
Neutral alumina	1
Acidic alumina	0.8
Basic alumina	0.3

Figure 1.3 Possible reactions of cyanides with acidic supports.

supported reagents (Table 1.4). The authors postulated that the surface acidity affects the ring opening directly (i.e., by interaction with the epoxide) rather than by an alteration in the inherent activity of the support NaN_3 itself.

Surface areas also vary dramatically from support to support—for example, CaF_2 is typically 10–15 m^2/g, Al_2O_3 100–200 m^2/g and SiO_2 500–600 m^2/g. Materials, such as clays (e.g., montmorillonites, hectorites) and graphite, are more complex—being lamellar materials they have an external surface area, but if the layers can be separated, the internal surface area also become available and this can be much more extensive than the external surface area—and is consequently often the more important of the two. For clays, such as montmorillonites, this separation is easily achieved by most solvents, and therefore, normal supporting methods produce interlamellar-supported reagents with all but the largest molecules. For others, such as micas, separation is less easy and the internal surface area is usually less accessible. Care must be taken to ensure that the material prepared is actually interlamellar. Two methods have been published for the reaction of CrO_3 with graphite—one produces a supported reagent based on external surface area and the other is intercalated. Evidence exists showing that the two materials are chemically different.

Croft[56] prepared a material thought to be interlamellar CrO_3–C that was shown by Lalancette et al.[57] to oxidize alcohols to aldehydes in high yields. Ebert et al.[58] showed that this material was actually externally supported Cr_3O_8–C. Truly intercalated CrO_3–C has been prepared by Platzer and de la Martiniere[59] and was shown to be devoid of any oxidative properties[60] (Fig. 1.4):

Table 1.4 Effect of Acidity on Ring Opening of Epoxides

Supported Reagent	Acidity (H_o)	Yield of 1 (%)	Yield of 2 (%)
NaN_3/CaY	$\leqslant -8.2$	90	7
NaN_3/NaY	0.8–1.5	49	12
NaN_3/KY	2.0–3.3	7	0.5

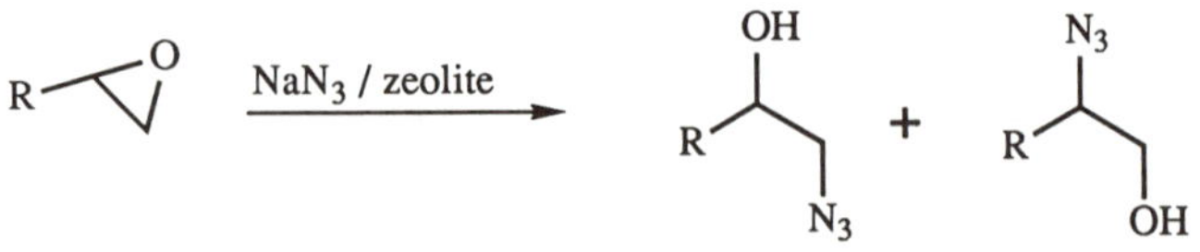

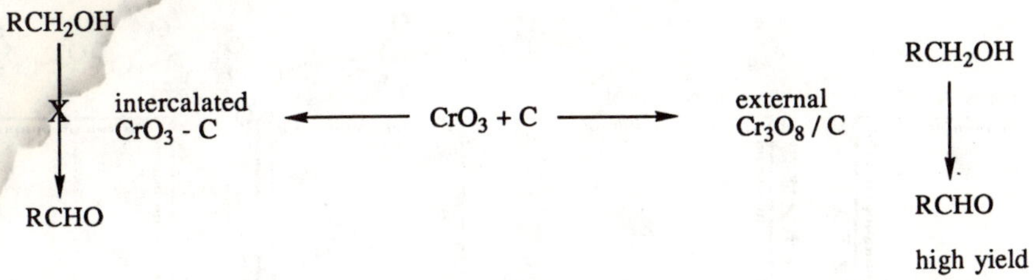

Figure 1.4 Variation of reactivity of intercalated versus externally supported reagent.

Zeolites have a rigid three-dimensional aluminosilicate structure containing channels, and therefore, do not require swelling (i.e., the opening of layers). The restricted size of the channels allows some very selective reactions to take place within the channels, but precludes the insertion of some reagents. Thus, in the case of $LiBH_4$ zeolites,[54] all the $LiBH_4$ is supported on the external surface of the zeolites.

Recently pillared clays have been prepared and used. These materials are, in effect, interlamellar ion-exchanged clays where the cation exchanged into the clay is a large cluster ion that then stops the clay layers from collapsing. They can be viewed (very approximately) as being somewhere between clays and zeolites.

It should be noted that dramatic changes in surface area within one support type may be accompanied by changes in the surface chemistry (and possibly the nature of the surface porosity). For example Clark and Macquarrie[61] found that a high surface area alumina (~230 m^2/g) chemisorbed only about 25% of the sodium dichromate that normal surface area alumina (120–150 m^2/g) would chemisorb. This reflects that either the number of appropriate surface sites is fewer in the high surface area alumina (changes in surface chemistry) or that the proportion of surface area accessible to the relatively large dichromate ion is *less* in the high surface area material (changes in porosity). It is important to note that surface area measurements are usually measured by N_2 adsorption. It is, therefore, possible that a larger amount of a surface may be accessible to nitrogen than to the much larger dichromate ion. Thus, even if the surface area accessible to nitrogen has increased, the surface area accessible to larger molecules–ions may have decreased. An idealized model is shown in Figure 1.5. It should be noted that this argument implies that the surface area of the supported reagent may be considerably different to that of the support. For reactions of the supported reagent this may often be a more relevant figure than the surface area of the support and thus the surface area of the support represents the maximum surface area that the supported reagent could have. How close supported reagents get to this value is not known.

Thus, if the high surface area alumina used has a high proportion of

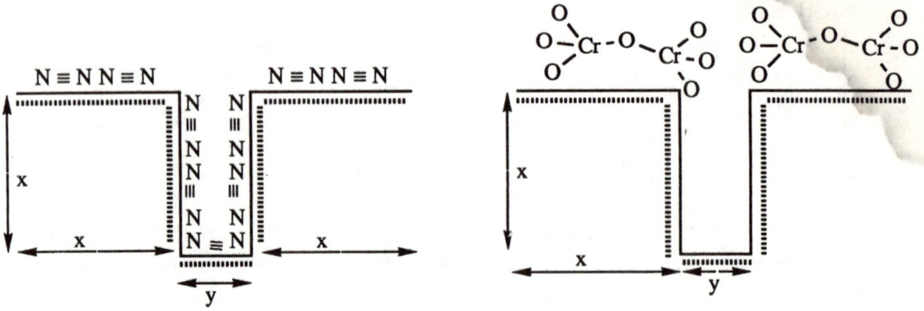

a) **Surface area accessible to N_2 = total surface area = 4x + y**

b) **Surface area accessible to $Cr_2O_7^{2-}$ ~ 2x < total surface area. Large $Cr_2O_7^{2-}$ ions cannot penetrate the pore**

Figure 1.5 Variation in available surface area with size of adsorbed species.

its surface area in narrow pores, this logic would imply that the available surface area of the high surface area alumina could be less than that of the normal variety. This can be illustrated by Figure 1.6. Unfortunately very little is known about such changes, very few surface area measurements having been made on the supported reagents themselves in spite of it being a common technique to characterize supports.

One paper details the change in surface area on preparing NaOH–SiO_2 at several loadings.[62] The surface area measurements were carried out at three loadings up to 0.9 mmol/g—still not a high loading for SiO_2—and showed a significant reduction in surface area as the loading increased (Fig. 1.7).

Figure 1.6 Possible changes in surface area due to variations in porosity.

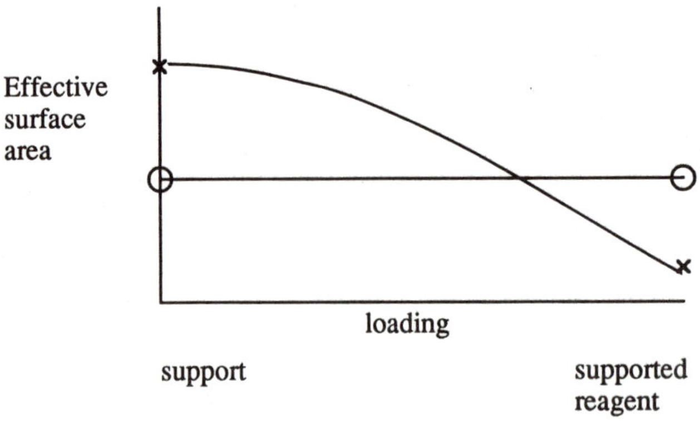

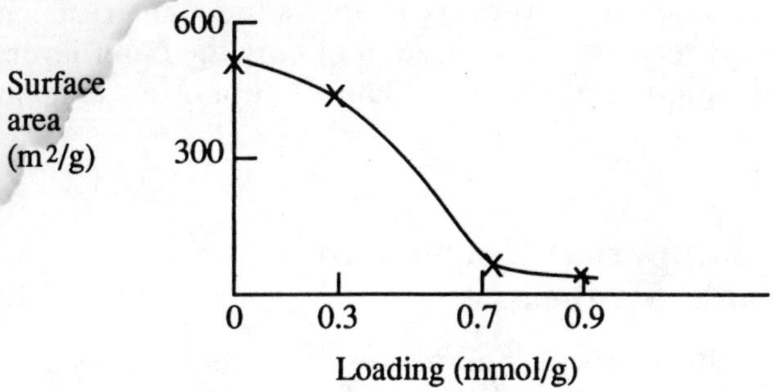

Figure 1.7 Variation of surface area of $NaOH–SiO_2$ with loading.

Further discussion on surface area measurements can be found in Chapter 3.

Supports are normally available in a number of physical forms—the usual variation being the particle size—typically measured as a mesh size. Generally speaking, mesh size has no effect on supported reagent properties—several investigators have used supports of differing mesh size and have found no differences between supported reagents prepared from them.

Some supports can also be obtained as pellets. These can behave very differently from the powdered support. For instance, Regen and Koteel[63] found that for potassium permanganate supported on various materials, pelletized supports gave supported reagents with lower oxidizing capacity than powdered supports. Owing to the difficulties in separating unsupported $KMnO_4$ from powdered-supported reagents, the loading of these materials was not calculated. Comparisons of oxidizing ability are given:

Supported reagent	*Oxidizing capacity* (mmol alcohol/g reagent)	
	Pellet	Powder
$KMnO_4$/LMS 13X	0.23	0.56
$KMnO_4$/K-306 montmorillonite	0.06	0.34

Two factors may influence this behavior. First, most materials are not capable of forming stable pellets on their own. Most require substantial amounts of binders—often clays—possibly as much as 30–40% by weight. Thus pelletized materials are usually chemically different from unpelletized materials.

Second, pellets are normally formed under pressure, and are tightly packed materials. Diffusion through the pellet can, therefore, be slow. This results in regions where no reagent has penetrated, leading to a smaller amount of supported reagent being formed. More important, the supported reagent formed may be seriously overloaded.

Similarly, diffusion of reactants–products in the subsequent reaction may also be very slow, and result in the utilization of only the outer layers of the pellet, again reducing the effective amount of supported reagent present.

1.4. Supported Reagents in Organic Synthesis

Appendix 1 provides numerous examples of the applications of supported reagents in organic synthesis. It has broad categories:

1. Oxidation and reduction
2. Addition to C-heteroatom multiple bonds
3. Addition to C–C multiple bonds
4. Intramolecular rearrangement
5. Aromatic substitution
6. Elimination
7. Aliphatic substitution

Appendix 1 is not intended to be an exhaustive literature search, but rather to give examples of the variety of reactions and supported reagents.

REFERENCES

1. S. A. McKillop and D. W. Young, Synthesis, (1979), 401 and 481.
2. J. H. Clark, A. P. Kybett, D. J. Macquarrie, S. J. Barlow and P. Landon, J. Chem. Soc., Chem. Commun., (1989), 1353.
3. A. Cornelis and P. Laszlo, Synthesis, (1985), 909.
4. N. Ishikawa, T. Kitazume, T. Yamazaki, Y. Mochida and T. Tatsuno, (1981), Chem. Lett., 761.
5. P. Laszlo, Acc. Chem. Res., (1986), **19,** 121.
6. T. Ando, J. H. Clark, D. G. Cork and T. Kimura, Bull. Chem. Soc. Jpn., (1986), **59,** 3281.
7. R. S. Drago and E. E. Getty, J. Am. Chem. Soc., (1989), **110,** 3311.
8. M. Hiro and K. Arata, Chem. Lett., (1989), 971.
9. H. Ogawa, N. Hiraga, T. Chihara, S. Teratani and K. Taya, Bull. Chem. Soc. Jpn., (1988), **61,** 2383.
10. H. Ogawa, T. Hagiwari, T. Chihara, S. Teratani and K. Taya, Bull. Chem. Soc. Jpn., (1988), **61,** 4371.
11. J. M. Adams, K. Martin, R. W. McCabe and S. Murray, Clays, Clay Miner., (1986), **34,** 597.
13. P. Laszlo and A. Mathy, Helv. Chim. Acta, (1987), **70,** 577.

14. G. Perot and M. Guisnet, J. Mol. Catal., (1990), **61,** 173.
15. A. Corma, H. Garcia and J. Primo, J. Chem. Res. (S), (1988), 40.
16. M. J. Climent, A. Corma, H. Garcia and J. Primo, App. Catal., (1989), **51,** 113.
17. K. Smith and M. Butters, Synthesis, (1985), 1157.
18. H. Molinari, F. Montanari and P. Tundo, J. Chem. Soc., Chem. Commun., (1977), 639.
19. V. Janout, J. Kahovec, H. Hrudkova, F. Svec and P. Cefelin, Polym. Bull., (1984), **11,** 215.
20. K. Sukata, Bull. Chem. Soc. Jpn., (1987), **60,** 1085.
21. A. K. Sinhababu and R. T. Borchardt, Tetrahedron Lett., (1983), **24,** 227.
22. J. H. Clark, A. P. Kybett, P. Landon, D. J. Macquarrie and K. Martin, J. Chem. Soc., Chem. Commun., (1989), 1355.
23. W. A. Smit, S. L. Kireev, O. M. Nefedov and V. A. Tarasov, Tetrahedron Lett., (1989), **30,** 4021.
24. G. H. Posner, Angew. Chem. Int. Ed. Engl., (1978), **17,** 487.
25. P. Laszlo, ed., Preparative Chemistry Using Supported Reagents, Academic, San Diego, (1987).
26. G. Bram, E. d'Incan and A. Loupy, Nouv. J. Chim., (1982), **6,** 689.
27. A. P. Kybett, Friedel–Crafts and Oxidation Catalysis using Supported Reagents, D. Phil. Thesis, York. (1988).
28. T. Ando, S. J. Brown, J. H. Clark, D. G. Cork, T. Hanafusa, J. Ichihara, J. M. Miller and M. S. Robertson, J. Chem. Soc., Perkin Trans. 2, (1986), 1133.
29. R. K. Iler, The Chemistry of Silica, (1979), Wiley, New York.
30. Surface area information has been obtained from Rhone-Poulenc Minerale Fine (alumine) and Sud-Chemie (montmorillonite).
31. K. H. Radeke, Carbon, (1984), **22,** 473.
32. J. A. Martens, J. Perez-Pariente and P. A. Jacobs in Chem. React. Org. Inorg. Constrained Syst., R. Setton, ed., NATO ASI Ser., Ser. C., (1986), **165,** 115–129.
33. K. Suzuki, T. Mori, K. Kawase, H. Sakami and S. Iida, J. Chem. Soc., Chem. Commun., (1988), 122.
34. K. T. Liu and Y. C. Tong, J. Org. Chem., (1978), **43,** 2717.
35. J. M. Melot, F. Texier-Boullet and A. Foucaud, Tetrahedron Lett., (1986), **27,** 493.
36. S. Quici and S. L. Regen, J. Org. Chem., (1979), **44,** 3436.
37. G. H. Posner and G. M. Gurria, J. Org. Chem., (1976), **41,** 578.
38. G. H. Posner, G. M. Gurria and K. A. Babiak, J. Org. Chem., (1977), **42,** 3173.
39. F. A. Lamb, P. N. Cote, B. Slutsky and B. M. Vittimberga, J. Org. Chem., (1974), **39,** 2796.
40. D. Villemin, J. Chem. Soc., Chem. Commun., (1983), 1092.
41. J. V. Sinisterra and J. M. Marinas, Tetrahedron Lett., (1986), **27,** 4971.
42. T. Ando, J. H. Clark, D. G. Cork and T. Kimura, Chem. Lett., (1987), 1315.

43. E. Santaniello and P. Ferraboschi, Nouv. J. Chim., (1980), **4,** 279.
44. S. L. Regen, S. Quici and S. J. Liaw, J. Org. Chem., (1979), **44,** 2029.
45. S. L. Regen and C. Koteel, J. Am. Chem. Soc., (1977), **99,** 3837.
46. H. Ogawa, Y. Ichimura, T. Chihara, S. Teratani and K. Taya, Bull. Chem. Soc. Jpn., (1986), **59,** 2481.
47. M. Kodomari, T. Kuzuoka and S. Yoshitomi, Synthesis, (1983), 141.
48. M. Kodomari, S. Takahashi, H. Watanabe and S. Yoshitomi, Nipp. Kag. Kai, (1987), 1753.
49. T. Funabiki, T. Sugimoto and S. Yoshida, Chem. Lett., (1982), 1097.
50. T. Ando, J. H. Clark, D. G. Cork, M. Fujita and T. Kimura, J. Org. Chem., (1987), **52,** 681.
51. T. Ando, J. H. Clark, D. G. Cork, M. Fujita and T. Kimura, J. Chem. Soc., Chem. Commun., (1987), 1301.
52. J. A. Ballantine, J. H. Purnell, M. Rayanakorn and K. J. Williams, J. Mol. Catal., (1985), **30,** 373.
53. J. H. Clark, A. P. Kybett, P. Landon, D. J. Macquarrie and K. Martin, J. Chem. Soc., Chem. Commun., (1989), 1355.
54. P. A. Risbood and D. M. Ruthven, J. Org. Chem, (1979), **44,** 3969.
55. M. Onaka, K. Sugita and Y. Izumi, J. Org. Chem., (1989) **54,** 1117.
56. R. C. Croft, Aust. J. Chem., (1956), **9,** 201.
57. J-M. Lalancette, G. Rollin and P. Dumas, Can. J. Chem., (1972), **50,** 3058.
58. L. B. Ebert, R. A. Huggins and J. I. Brauman, Carbon, (1974), **12,** 199.
59. N. Platzer and B. de la Martiniere, Bull. Soc. Chem., (1961), 177.
60. J. P. Alazard, H. B. Kagan and R. Setton, Unpublished Results.
61. J. H. Clark and D. J. Macquarrie, Unpublished Results.
62. A. W. Klaason and C. G. Hill, J. Catalysis, (1981), **69,** 299.
63. S. L. Regen and C. Koteel, J. Am. Chem. Soc., (1977), **99,** 3837.
64. J. M. Thomas, Sci. Amer., (1992), 82.

2 *Preparation of Supported Reagents*

The aims of preparing a supported reagent are to obtain a material of optimum efficiency with regard to yield and rate, and of optimum selectivity (i.e., the promotion of the desired reaction path with the elimination of side reactions). Optimization of these factors should also provide the additional benefit of an easier workup procedure at the end of the reaction.

Various methods have been developed to prepare supported reagents, each having their own strengths and weaknesses. To determine which method is most appropriate in a particular case, several factors must be taken into account. These factors are:

1. Choice of support (described in Chapter 1)
2. Support pretreatment—drying, wetting, modification of surface, and so on
3. Choice of loading—ratio of reagent to support, normally expressed in millimole of reagent/gram support
4. Choice of solvent
5. Method of preparation
6. Supported reagent posttreatment—further reaction of surface-bound species, drying, and so on

Although it is impossible to provide a definitive set of rules with which to determine the "correct" choice, a discussion of the preceding factors should provide guidance in determining the details of the preparation.

2.1. Support Pretreatments

Supports are usually used from the bottle. In some cases, however, it is advantageous to pretreat supports before the preparation of the supported reagent. Most of these examples are concerned with drying the support; the supported reagent may be less active when wet or may need a specified amount of water to display optimum activity. These examples are discussed under the sections on posttreatment, as control of the water content of supported reagents is best achieved through posttreatment. Therefore, the discussion of pretreatment is restricted to those cases in which the failure to pretreat the support results in a less active supported reagent system that cannot be rectified by posttreatment. Again, most examples deal with water content, but in a different context. Several preparations of supported reagents are affected by the presence of water because of the nature of the reagent being supported. It is also possible to affect the surface sites by pretreatment, thereby increasing the number of sites with the required activity.

2.1.1. Support Drying

Montmorillonites can be dried under vacuum by heating to ~120°C. This treatment removes the interlamellar water from the clay. Choudary and co-workers [1,2] have used such dehydrated clays for the specific synthesis of anchored palladium catalysts (see Section 2.2.7.2). During these syntheses, reagents were used (such as Ph_2P-Cl) that are hydrolytically unstable. Use of vacuum-dried montmorillonite allowed these materials to be used efficiently in the preparation of various clay-bound catalysts.

Similarly, the preparation of $AlCl_3$–Al_2O_3 superacids was preceded by the formation of dehydrated Al_2O_3 at 550°C to avoid hydrolysis of $AlCl_3$ by water.[3] Preparation of superbasic materials based on alumina has been carried out using the condensation of NaOH and then Na onto precalcined (500°C) alumina with removal of water formed.[4] The importance of pretreatment was not discussed.

The effect of thermal pretreatment of silica on the CsCl–SiO_2 dehydrochlorination of 1,1,2-trichloroethane has been investigated[5] (Fig. 2.1). The silica was pretreated for 2 h at temperatures of 100–600°C. The surface area decreased over this temperature range starting at 650 m^2/g. No change was observed until 400°C when a gradual reduction began;

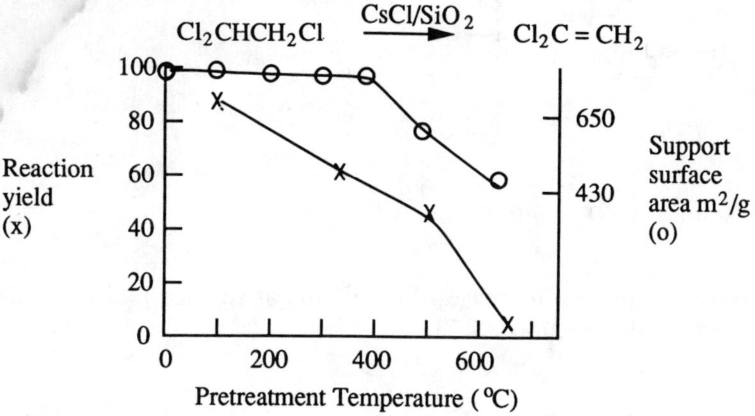

Figure 2.1 Effect of pretreatment temperature on the activity of CsCl–SiO_2.

after treatment at 600°C, the surface area was 430 m^2/g. Over this range of temperatures, the activity of the supported reagent was also measured. Supported reagents prepared from pretreated silicas showed a marked decrease in activity over the entire range of pretreatment temperatures starting at 100°C, from 92% to 12%. Thus, the effect of reducing surface area is not the only factor that changes with calcination, and other factors are important in the determination of activity of the supported reagent. It should be noted that the surface area of the supported reagent itself was not measured.

The elimination of unwanted side reactions is another aspect of pretreatment. Although supported reagents can normally be posttreated to avoid side reactions, in situ reagents cannot, and therefore, pretreatment of the support, and often the reagent, must be carried out. An example is the use of alumina in the elimination of tosylates to give alkenes.[6] Although this is not a supported reagent reaction, it illustrates the point. Hydrolysis is a problem when this system is applied to acyclic tosylates. The hydrolysis is caused by water bound to the alumina. Drying of the alumina at 400°C/0.006 mmHg dramatically reduces the amount of hydrolysis. The conditions used are sufficient to remove nearly all the bound water without causing damage to the surface.

2.1.2. Hydrolysis of Surface Groups

By contrast, Tundo and Venturelli[7,8] have pretreated silica by refluxing in HCl. This is a procedure that hydrolyzes Si-O-Si units[9] (Fig. 2.2). The pretreated silica was then used to prepare silica-bound phase transfer catalysts by reaction with the surface hydroxyls (Fig.

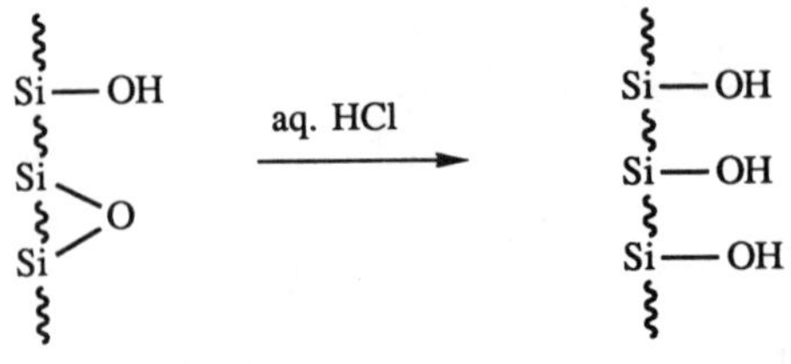

Figure 2.2 Reaction of SiO_2 with HCl.

2.3). A pretreatment that increases the amount of surface hydroxyls will be beneficial (see also Section 2.2.7).

2.2. Methods of Preparation

A variety of procedures to prepare supported reagents can be found in the literature. The use of these methods has led to supported reagents with many different applications. It is possible to support an enormous variety of substances, ranging from highly reactive, aggressive molecules, such as anhydrous ferric nitrate, to simple salts such as sodium acetate and cesium chloride.

This section details these methods and attempts to illustrate their usefulness and limitations. The information given here will hopefully assist the reader in deciding which route to take to a given reagent.

In the literature, the preparation methods are divided, fairly arbitrarily in some cases, into 10 types. Arguably, some supported reagents could be in different sections, and some may be considered as being in more than one, for example, a specific synthesis of more than one step followed by subsequent posttreatment. It is hoped that, despite this possible objection, the following classification will form a useful practical basis for the rational design and preparation of supported reagents.

The following division of preparative methods has been made and is used in this chapter:

1. Evaporation
2. Precipitation
3. Adsorption from solution

Figure 2.3 Formation of surface bound reagent.

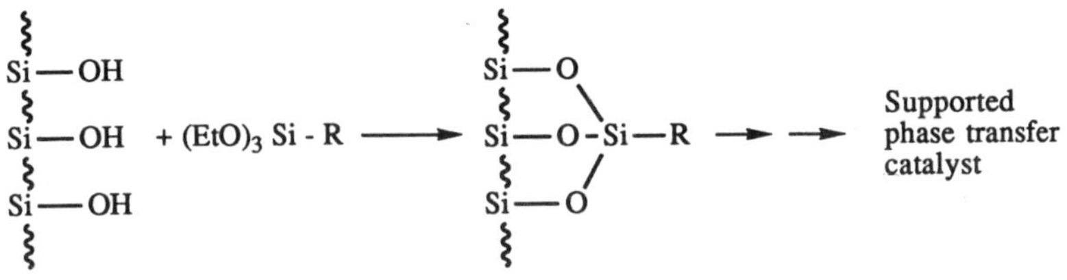

4. Mixing/grinding
5. In situ
6. Ultrasound
7. Specific syntheses
8. Substrate adsorption
9. Ion exchange

These nine methods cover essentially all the methods used so far. Other methods may yet be developed and used, which will give access to different types of supported reagents. Some of the techniques are widely applicable, others very specialized, some have hardly been explored but are of interest and may become much more widely used and understood.

The remainder of this section consists of a summary of each technique's advantages and disadvantages, followed by more detailed examples of the use of each technique.

Method	Advantages	Disadvantages
Evaporation	Very widely applicable—most common method used. It is relatively easy and quick to use and gives the greatest control over the evenness and level of loading.	Need of an appropriate, relatively volatile solvent for preparation—not always available. Not ideal method for some unstable reagents.
Precipitation	Allows preparation of supported reagents from sparingly soluble/insoluble materials without the use of excessive solvent volumes/heating, or the necessity of volatile solvent systems.	Best variant requires reaction to produce insoluble reagent. Recrystallization of reagent in presence of support unknown, may be too difficult to control as support may act as seed. Technique is best if precipitation is controlled and slow.
Adsorption from solution	Reaction specific to certain sites on surface. This may produce unusual reactivity of supported reagent due to chemisorption. Generally mild conditions and simple to carry out.	Not general or widely applicable. Almost no control over surface coverage except, to some extent, by varying support. Large amounts of exposed surface may cause side reactions. Large amount of support often required for a given amount of reagent.

Method	Advantages	Disadvantages
Mixing/Grinding	Simple to perform. Can be very efficient in some cases—no need for solvents or heating, thus insoluble unstable reagents can be used.	Often does not produce supported reagent. May produce less than optimum supported reagent owing to incomplete coverage of surface, incompletely supported material, etc. In these cases, it is better to use other methods of preparation.
In situ	No actual preparation necessary. Again can be very effective in some cases.	Same as for Mixing/Grinding
Ultrasound	No actual preparation required. It appears that ultrasound can activate the supported reagent as well as assist in its preparation.	Little known about scope of reaction or of the effect of ultrasound on the supports normally used.
Specific syntheses	Provides method for complex supported reagents, often impossible to prepare in other ways. Reagents prepared are often better defined than in other methods.	More complex than other methods. Can sometimes require long synthetic routes. Not a general method as it relies heavily on the surface chemistry of the support.
Substrate adsorption	Can give very unusual, and often very high, selectivities.	Need *ordered* monolayer adsorption on surface. Limited to substrates that will provide this feature, and also to reaction systems that will maintain it. This leads to restrictions on the solvent and the reagent.
Ion exchange	Reactivity of ion-exchanged materials may be complementary to supported equivalent. Different materials are produced in each case.	Limited mainly to clays, although silica has also been used. Often this method is slow and excess of reagent is required for complete exhanges.

It should be remembered that the preceding summary is a general analysis of the pros and cons of the techniques, and that exceptions always exist. A more detailed survey of the techniques, along with examples of their use, where appropriate, is now given.

2.2.1. *Evaporation*

The most common method of preparing supported reagents is the evaporative technique. This technique is suitable for a wide range of reagent/support combinations, and is usually the method of choice, allowing exact control of the loading, and also ensuring thorough mixing of, and contact between, the support and the reagent.

Such preparations are carried out on a rotary evaporator, which is very well suited for such an operation.

A typical procedure involves the following steps:

1. Dissolve the reagent in a suitable solvent in a rotary evaporator flask. For practical reasons, solvents should be volatile enough to be easily evaporated on a rotary evaporator; however, it is possible to use solvents, such as *N*-methyl pyrollidone (b.p. 200°C), if necessary. In this case, removal of the last traces of the solvent can be difficult. Typically concentrations are of the order 0.1–1 M, although this is more for convenience than for any fundamental reason. The solution should be substantially below saturation concentrations, and should be of sufficient volume to thoroughly mix with and wet the support when it is added.
2. An appropriate amount of support is then added to the solution. The mixture is then swirled and stirred on the rotary evaporator to achieve uniform wetting and mixing of the solid and the solution.
3. The solvent should be removed slowly. Even mixing is very important at this stage for the production of an optimum-supported reagent. This is best achieved by slow rotation of the flask. It stops the centrifugation effects that can force the solids against the walls of the flask and hinder proper mixing. Slow rotation also ensures that all the solid remains immersed in the solution for as long as possible. This is important because during the process of evaporation, the concentration increases and the reagent becomes increasingly adsorbed onto the solid surface. Any solid not constantly immersed in the solution will not adsorb as much as a solid that is in the solution all of the time. This leads to some solid being underloaded, some being overloaded.
4. As the solution evaporates, the solid often becomes a damp mass and eventually a solid crust. Continued evaporation of the solvent produces a dry, free-flowing powder, usually of the same texture as the support. This may require a slight increase in bath temperature.

5. The supported reagent is now ready for posttreatment/use, and so on. It should not contain lumps or aggregates. Nor should it, if properly loaded, appear heterogeneous. The easiest to check are colored reagents supported on white solids. A well-prepared supported reagent will be of uniform color, and if the loading is correct it will contain no free crystals.

This procedure can be applied to a wide range of supports and reagents. With appropriate care, it can be used for unstable supported reagent preparations. One such example is clayfen, prepared by Laszlo's group. This material is ferric nitrate supported on K10 montmorillonite.[10] Under the conditions of the reaction the actual species deposited is closely related to the acetone solvate of ferric nitrate, a highly unstable oil that decomposes vigorously. It is possible, with care and close observance of the correct procedure,[10] to safely produce a highly reactive, supported reagent in a relatively stable form. Deviation from the procedure can produce a reagent that can decompose vigorously with the formation of large quantities of nitrogen dioxide.

Of interest in this preparation is the fact that for a stable reagent to be formed, the dry crust must be flaked off and crushed, then further evaporated to a powder. Normally, this dry crust can be left to dry, as it is, to a powder without any adverse effect on the final reagent. Presumably, in the case of clayfen, the flaking is done to speed up the evaporation, thus avoiding heating the material longer than necessary.

This method of preparation is versatile and can be used to prepare a wide range of supported reagents. It allows complete control over the loading, which is determined by the initial ratio of reagent to support. This procedure does not distinguish between chemisorbed and physisorbed material, however, allowing both types to be formed where possible. This is rarely a problem, and if the active species is the chemisorbed material, then washing will remove the physisorbed material. Intercalated material can be prepared in this way, such as clayfen and MX_n/K10,[11] provided the solvent is capable of opening up the clay layers.

2.2.2. *Precipitation*

Precipitation of inorganic salts onto supports is an effective method of preparing supported reagents, mainly where solubility is a problem, in which case the reagent can be prepared by reaction of a soluble salt to form an insoluble product in a well-stirred suspension of support. Indeed, the first reported supported reagent, silver carbonate on celite, was prepared in such a manner by Fetizon and Golfier.[12] Several methods have been used to prepare supported reagents by precipitation, and it seems that most methods of precipitation, such as reaction, cooling, or addition of cosolvents, can produce well dispersed supported reagents.

In general, preparations of this type should be carried out in well-agitated reactors, ensuring that all the support is suspended. Also, best results are obtained if the precipitation is gradual, rather than rapid.

Examples of different methods of precipitation are now given

2.2.2.1. Precipitation By Cooling. Copper(I) salts are fairly insoluble in common organic solvents. However, they are quite soluble in hot NMP. Thus, Clark and Jones.[13] used hot NMP as a solvent for CuCN in preparation of supported CuCN (on Al_2O_3, SiO_2, and charcoal). The high boiling point of NMP precludes efficient removal by evaporation (indeed the active form of the supported reagent is unstable much above 100°C) and therefore, the supported reagents were prepared by slow cooling of a well-agitated slurry of the support in a hot solution of CuCN in NMP. This proved to be an efficient method to produce dispersed supported reagents of high activity. After washing the filtered reagent with ether (to remove NMP) the material was analyzed and shown to have a well-dispersed monolayer of precipitated salt. As in all methods that involve removal of a solvent, the filtrate was checked for dissolved CuCN; none was present. X-ray diffraction showed no crystalline CuCN on the supported reagent. Thus, by choosing the correct amount of CuCN for monolayer coverage (~5 molecules/nm^2) a monolayer coverage was achieved with no microcrystalline material formed.

2.2.2.2. Precipitation by Adduct Decomposition. In another preparation of supported Cu(I) salts, Clark and Jones[13] have used the equilibrium formation of $KCuI_2$ as a method to support CuI. A standard method to purify CuI is to dissolve it in concentrated KI solution, thus forming the soluble $KCuI_2$:

$$KI_{(aq)} + CuI \longrightarrow KCuI_2$$

Addition of water to this solution reforms the purified CuI as a precipitate, which can be collected and washed.

Carrying out this procedure in the presence of the desired amount of support, again with vigorous stirring and a very slow addition of water, gave highly active supported reagents. CuSCN/C has been made by the same type of reaction.[14] Slow decomposition gave an active reagent, whereas fast decomposition gave an inactive reagent.

2.2.2.3. Precipitation by Reaction. The first supported reagent, Ag_2CO_3–celite, was prepared by a precipitation reaction[12]:

$$2AgNO_3 + Na_2CO_3 \longrightarrow Ag_2CO_3 + 2NaNO_3$$

Carrying out this reaction in the presence of celite gave the supported reagent, which can be filtered off and washed before azeotropic drying.

Another example of a precipitation reaction is the preparation of AgF–

CaF_2.[15] In this preparation, the reaction is carried out as an evaporative reaction. Thus, silver carbonate and aqueous HF are mixed with CaF_2, and the slurry is slowly evaporated to dryness. The following scheme illustrates the reaction:

$$Ag_2CO_3 + 2HF \xrightarrow{CaF_2} AgF/CaF_2 + H_2O + CO_2$$

The reagent is a highly active source of nucleophilic fluorine that is more easily handled than silver fluoride itself.

2.2.3. *Adsorption from Solution*

Supported reagents can be prepared by simply stirring, or refluxing, a solution of reagent in the presence of the support material. Because this requires a relatively strong support–reagent interaction, it can be implied that some form of chemisorption via reaction with the surface takes place. In several cases, this has been shown to be true.

Such a supported reagent is prepared by simply stirring/refluxing a solution containing the reagent in the presence of an appropriate amount of support. Filtration and washing of the solid, followed by any relevant posttreatment (e.g., drying) gives the supported reagent. Several points should be made regarding this process.

1. In contrast to an evaporative preparation where all the reagent is forced to adsorb onto the surface, there is no guarantee that all the reagent will adsorb in this type of preparation. Thus, the loading must be calculated, either directly, or more commonly (and simply), by analysis of the supernatant liquid using conventional analytical techniques. This is important for two reasons:
   (a) For stoichiometric reactions, knowledge of the amount of reagent adsorbed per gram of supported reagent allows sufficient supported reagent to be added, and helps in determining reaction stoichiometry.
   (b) A knowledge of loading, compared with the monolayer loading capacity of the support, determines the amount of exposed surface present. This is important as many supports have more than one type of site. Chemisorption may involve only one type, leaving others exposed, possibly causing side reactions, although in some cases it may be that these exposed sites on the support could be beneficial. Because supports differ considerably in the nature and amount of active sites, this information is important in choosing the correct type of support and the correct variant of that support (e.g., acidic, neutral, or basic alumina—high surface area or low—for the particular application).

2. The nature of the reaction(s) occurring during the preparation of the supported reagent can be looked at by analysis of by-products. This is often useful in determining the chemisorption reaction stoichiometry (along with details of unreacted reagent) and reaction products. It may also be possible to infer the occurrence of two types of reaction by analysis of by-products/unreacted reagents.
3. Knowledge of all these factors can lead to a reasonably detailed understanding of the structure of the supported reagent, which may help rationalize its reactivity, and could be crucial in the development of improved versions.
4. Knowledge of the extent and nature of chemisorption can also be valuable in the study of supported reagents prepared by evaporation. Some of these contain a proportion of chemisorbed material, the rest being physisorbed, often as microcrystals. Comparison of the chemisorbed with the physisorbed material is thus possible by comparing the reactivity of both types of supported reagent. Such a study can be very important in the understanding of supported reagents in general.

These points are illustrated in the following examples:

1. Chromyl chloride can be adsorbed onto silica–alumina by stirring a methylene chloride solution of CrO_2Cl_2 in the presence of the support material[16] (Fig. 2.4). The adsorption reaction can be followed by the loss of color from the solution onto the support. The loading of chromium on the reagent is, therefore, known. No mention is made of by-products, such as HCl, that could arise from the reaction with surface hydroxyls. The absence of any evolved HCl would indicate that a different surface species is present.
2. $LiBH_4$ can be supported on zeolites by refluxing a THF solution of $LiBH_4$ in the presence of zeolites.[17] This gives a reagent capable of selectively reducing aldehydes in the presence of ketones. Owing to the nature of $LiBH_4$, it was necessary to predry carefully all the reagents and solvents. Analysis of the absorption process was carried out by titration of hydrolyzed supernatant solution or by atomic absorption

Figure 2.4 Possible reactions of CrO_2Cl_2 with surface hydroxyls.

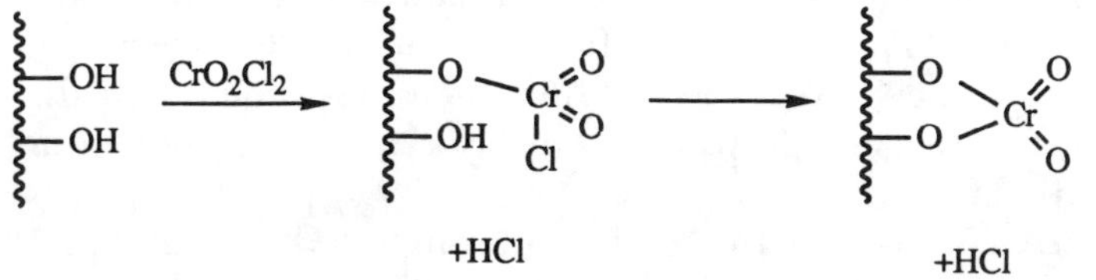

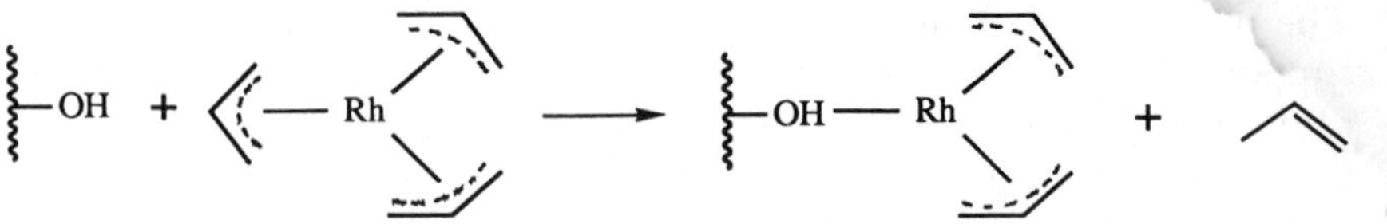

Figure 2.5 Reaction of *tris*-allylrhodium with silica surface hydroxyls.

spectrometry. This analysis provided important details about the nature of the reagent:

(a) The amount of $LiBH_4$ adsorbed by the sieves was relatively low and indicated that the $LiBH_4$ was only adsorbed on the surface of the zeolite, suggesting that reaction takes place on the surface of the zeolite, and not within the pores.

(b) This information let to mechanistic conclusions concerning the source of the selectivity of the reduction. The adsorption experiments proved that the selectivity did not arise from the fact that aldehydes, but not ketones, can penetrate the zeolite structures but must be due to other effects, as yet not known.

3. The preparation of silica-bound rhodium complexes has been described.[18] These were prepared by stirring a toluene slurry of dried silica and adding to it a solution of the yellow *tris*-allylrhodium (Fig. 2.5). The color of the solution disappears as the organometallic becomes adsorbed. The loss of 1 mole of propene indicates a specific surface reaction to create a surface-bound rhodium species.
   Such information is valuable to characterization, and formed the basis of a detailed description of the chemistry of the supported rhodium catalysts prepared in this work (see also Section 2.5.1).
4. Supported $AlCl_3$ catalysts are of great interest as solid acids/superacids. Krzywicki and Marczewski[3] have prepared a solid acid system by subliming $AlCl_3$ through a bed of dried (823 K) alumina, followed by evacuation to remove physisorbed $AlCl_3$. The resulting catalyst exhibits superacid properties of alkene isomerization. Analysis of the amount of Cl on the surface, as well as infrared data, suggest a complex structure.

   Another variation of strongly acidic-supported reagents has been prepared by Drago and Getty.[19] This time dry alumina (SiO_2, B_2O_3, or zeolites were also good) was reacted with $AlCl_3$ in refluxing CCl_4 (1.1 moles of HCl being evolved). Other solvents did not result in acidic catalysts. The loss of HCl suggests the formation of -$OAlCl_2$ species on the surface, with some O_2AlCl species. $AlCl_3$ does not sublime off the surface. Thus, the reagent prepared probably can be represented as shown in Figure 2.6.
5. The importance of checking the loading of supported reagents prepared by adsorption is illustrated by the reactions of $Na_2S–Al_2O_3$ re-

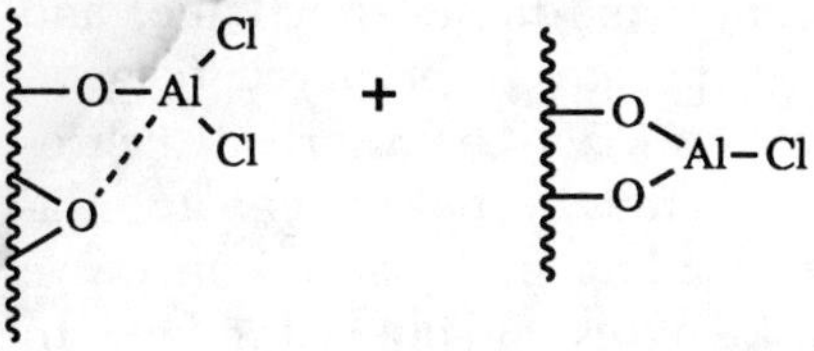

Figure 2.6 Possible structures for alumina bound $AlCl_x$ species.

agents. Kamitori et al.[20] reported that the reduction of aldehydes could be carried out using $Na_2S–Al_2O_3$. The reagent was prepared by mixing Al_2O_3 with aqueous Na_2S, followed by filtration and washing. This reagent was described as changing the selectivity of Na_2S reduction.

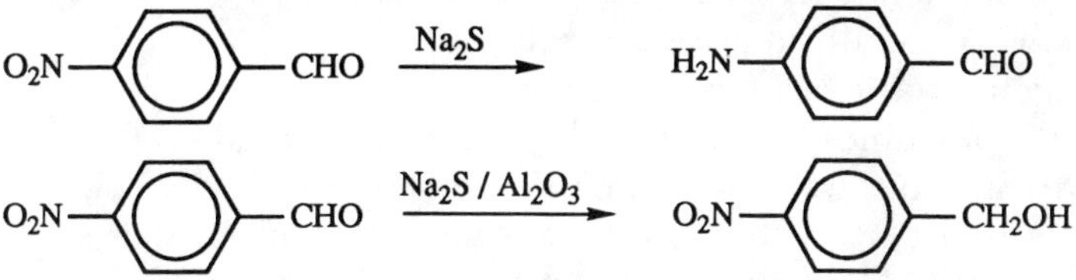

Work carried out by Ando et al.[21] showed that this conclusion was, in fact, wrong. They showed that Hojo's preparation resulted in only ca. 50% of the Na_2S being adsorbed on the alumina surface. This preparation gives a loading of 0.5 mmol/g, about one quarter of that required for monolayer coverage on alumina. Thus, a large proportion of the alumina surface is exposed. Ando showed that the "reduction" is actually a Cannizzaro reaction promoted by the exposed alumina surface. The acid products are always strongly bound to the alumina surface. Increasing the loading of Na_2S (by evaporative preparations) reduces the extent of reaction, stopping it altogether at slightly above monolayer coverage. They also showed that alumina itself promotes the reaction with no Na_2S present. Thus, knowledge of reagent loading is important if correct conclusions about the reactivity of the supported reagent are to be made.

6. A comparison of evaporative versus adsorption techniques has also been used to increase the understanding and effectiveness of supported reagents by Clark et al.[22] In this case the supported reagents are alumina-supported dichromates and permanganates. These materials are efficient oxidation catalysts, converting diphenylmethane to benzophenone with air as oxidant. The evaporatively prepared dichromate catalysts were loaded at 0.5–3 mmol/g and gave oxidation rates of 1% hr (w.r.t. substrate). Washing of the catalyst showed that 0.075 mmol/g dichromate was chemisorbed. The same loading can be achieved by stirring the alumina with a solution of dichromate in water, followed by filtra-

tion. This catalyst gives 1.3%/h oxidation of diphenylmethane, and therefore, is at least as active as the evaporatively prepared reagent. The catalytically active species in both cases is a chemisorbed dichromate species. Similar behavior was found for the permanganate catalysts. In this case, therefore, the lower loading gives more impressive results, with the physisorbed material being essentially redundant. In contrast to Example 5, exposed alumina surface has no detrimental effect on the reaction.

2.2.4. *Mixing/Grinding*

Supported reagents have been prepared by simply mixing or grinding together the support and the reagent in the absence of solvent. These materials are often used directly, although some require drying as a postpreparation stage. This represents an easy technique that can lead to some very useful supported reagents.

Different reagents have been prepared in this way.[23–26] In all cases, reactivity is good, sometimes better than the same reagent–support combination prepared in other ways. In some cases, it is obvious that chemical interaction is an important factor, in others it has been shown that such an interaction is absent. Thus, the success of this technique would appear to be dependent on various factors for different supported reagents. Some of these materials are discussed.

2.2.4.1. $KMnO_4$ on $CuSO_4 \cdot 5H_2O$ or bentonite.[24] $CuSO_4 \cdot 5H_2O$ and bentonite are both prepared by grinding in a mortar and are used with no posttreatment.

The selectivity of the reagent is considerably different from that of potassium permanganate itself, oxidizing double bonds only very slowly, and only if excess reagent is used.

$$C_6H_{13}{-}\overset{\overset{\displaystyle OH}{|}}{C}H{-}CH{=}CH_2 \xrightarrow[\text{or } KMnO_4/\text{bentonite}]{KMnO_4/CuSO_4 \cdot 5H_2O} C_6H_{13}{-}\overset{\overset{\displaystyle O}{\|}}{C}{-}CH{=}CH_2 \quad >80\%$$

$$PhCH{=}CH{-}\overset{\overset{\displaystyle OH}{|}}{C}H{-}CH_3 \xrightarrow[\text{or } KMnO_4/\text{bentonite}]{KMnO_4/CuSO_4 \cdot 5H_2O} PhCH{=}CH{-}\overset{\overset{\displaystyle O}{\|}}{C}CH_3 \quad >90\%$$

Potassium permanganate itself is highly reactive toward double bonds, being used as a chemical test for unsaturation.[27]

The reaction of free $KMnO_4$ with double bonds is so rapid that this behavior implies that free $KMnO_4$ is absent or only present in very small amounts. This, in turn, implies a rapid reaction between the permanganate and the supports. With the two supports being so different in character, it is not obvious what this reaction may be.

$$12KF + Al_2O_3 \longrightarrow 3K_2O + 2K_3AlF_6$$

$$3K_2O \xrightarrow{3CO_2} 3K_2CO_3$$

Figure 2.7 Reactions of KF with Al_2O_3.

Alternatively, free permanganate is present, and the change in selectivity is due to an interaction between the double bond and the support–supported reagent, resulting in a "protected" double bond that is less reactive than the alcohol function.

2.2.4.2. KF–Al_2O_3. Potassium fluoride supported on alumina is a widely used material, being an efficient base in a number of reactions. Several methods of preparation have been used to make KF–Al_2O_3. Texier-Boullet and Lequitte[25] have shown that simply mixing KF and alumina produces an active basic material. This material is of similar activity to that prepared from solution, and is typical of KF–Al_2O_3 rather than of KF or of Al_2O_3. It has been shown recently[28] that mixing KF and Al_2O_3 causes a facile reaction to take place (Fig. 2.7). Further details of this reagent system are given in Chapter 4.

Thus, the activity of this supported reagent derives from a reaction between the support and the reagent.

2.2.4.3. KOH–Al_2O_3. Regen and Mehrotra[23] have used KOH ground with alumina as a reagent for the hydrolysis of esters. Neither reagent, used separately after grinding, effected any hydrolysis. Preparing analogous supported reagents from aqueous KOH always produced less active reagents. Excessive drying of these regents resulted in complete loss of activity. These results show that a small but finite amount of water is necessary for an efficient reagent. Drying conditions were never severe enough to destroy the nature of the Al_2O_3 surface.

It is not possible for KOH to react with Al_2O_3 as it does with KF (the products are KOH and Al_2O_3!), although some surface reaction may be occurring. For example, KOH could react with Al-OH groups, converting them to K^+-OAl units with the generation of water. These strongly basic surface sites could act as a catalyst for the hydrolysis reaction. Because the reaction is a hydrolysis, water is necessary in stoichiometric amounts, explaining why excessive drying reduces activity. Too much water may retard the formation of the K^+-OAl centers and reduce the efficiency of the material. The alumina is predried, therefore, most of the water in the system comes from the KOH or from its reaction with the Al_2O_3 surface. The active water may be adsorbed onto the alumina surface close to the active site. Such a situation may be visualized as shown in Figure 2.8. Such

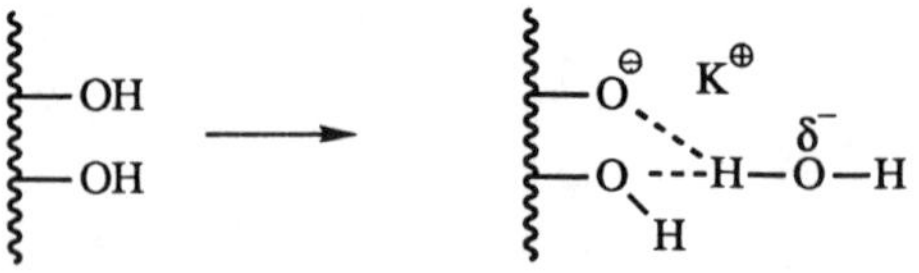

Figure 2.8 Possible interactions of KOH with an alumina surface.

a combination of sites may prove ideal for ester hydrolysis. Excessive drying would remove the H-bonded water from the surface, reducing the efficiency of the reagent.

2.2.4.4. Na–NaOH–Al$_2$O$_3$. A solid superbase ($H_- > 37$) has been prepared by Suzukamo et al.[4]. The first step is carried out by stirring NaOH and alumina at 310°C with the generation of water. A process similar to that postulated in Section 2.2.4.3 must be occurring, and indeed β-sodium aluminate was shown to be formed. This material is then treated with Na metal, with stirring, at 310°C to give a pale blue solid with superbasic properties. ^{23}Na NMR shows the presence of Na^+, but no Na(0). This is proof that the mixing procedure produces an active species by chemical reaction, and indicates that such "dry" reactions can be highly efficient.

2.2.4.5. KF–CaF$_2$. In contrast to the last four examples, KF–CaF_2 is a supported reagent in which there is no chemical interaction between the two components.[26] The reagent is prepared by mixing KF and CaF_2, grinding them together and drying at 150°C under vacuum. Powder, X-ray diffraction indicates no interaction between the two materials. This point has also been emphasized by Clark et al.[29] who studied KF–CaF_2 prepared from methanol by evaporation.

The activity of the ground KF–CaF_2 reagent in the fluorination of various halo aliphatics has been shown to be in excess of KF alone (CaF_2 is inert in these reactions). Despite the lack of chemical interaction, rate enhancements have been produced in this system. The observed rate enhancements are consistent with increasing the surface area of the KF. Such rate enhancements are known, for example, from normal KF (0.1 m^2/g) to spray-dried KF (1.3 m^2/g).[30] The rate enhancement of spray-dried KF to KF–CaF_2 prepared by evaporation was also explained by an increase in surface area[49] (surface area of $CaF_2 \sim 10\ m^2/g$).

Whether this process has actually coated the CaF_2 with a well-dispersed layer of KF (as in the evaporative reagent) or has simply acted as an abrasive powder to break down the KF particles more efficiently is not clear. Either method would produce a larger KF surface area, and thus, a more active reagent.

2.2.5. *In situ*

Of all the methods available for the preparation of supported reagents, the simplest is to add both support and reagent to the reaction mixture with no actual preparation step beforehand. Often this method does not work, either no reaction occurs or reactions due to the support itself take place, leading to different products to those intended. In some cases, however, supported reagents do seem to be formed, as indicated by changes in rate or reaction course. In some cases these can be compared with the same support–reagent combination prepared in different ways.

Apart from the fact that using supported reagents prepared in situ avoids having to carry out a separate preparation step, one other reason for using such a reagent should be mentioned. It may be difficult or impossible to prepare some supported reagents in any other way because of the instability of the reagent during the preparation step. Such instability may mean an inefficient reagent due to decomposition or the formation of an unreactive form of the reagent because of the conditions necessary to support it (care must be taken in the preparation of supported CuX salts for the latter reason). Some reagents are not stable and may be liable to decompose instantly during attempted preparation using normal techniques. (Some supports, on the other hand, actually stabilize unstable molecules, examples being clayfen–anhydrous $Fe(NO_3)_3$ supported on K10 and dynamite—trinitrotoluene/clay.)

Possible reasons for the changes in rate/reaction course of in situ supported reagents can be summarized in the following way (although other modes of action may also exist):

1. Insoluble reagents:
   (a) Mixing/grinding to form supported reagent (see Section 2.2.4).
   (b) Reaction of support with reagent to generate new surface-bound or soluble reagent.
2. Soluble reagents:
   (a) 1b above
   (b) Adsorption of reagent to form supported reagent (see Section 2.2.3).
   (c) Adsorption of substrate followed by reaction with reagent (see Section 2.2.8).
   (d) Changing reaction course by intercepting reaction intermediate from solution.

Some examples of these types of supported reagents are given now:

2.2.5.1. *KF–CaF*$_2$ *and CsF–CaF*$_2$[31].

These supported reagents are highly active sources of nucleophilic fluorine, and three methods have been used for their preparation.[30,31]

(a) Evaporative (from methanol)
(b) Mixing/grinding
(c) In situ

In the case of (a) and (b), it has been shown that the two components do not interact chemically. It is also known that reagent (b) is essentially as active as reagent (a), inferring that mixing/grinding in this case produces a well-dispersed, high surface area material. It appears that this process is occurring with the in situ route, again providing a supported reagent in the reaction system itself. The efficiency of the in situ supported reagent is significantly greater than the unsupported fluoride, but slightly less than the reagents (a) and (b). This implies that reagent (c) is less well dispersed than (a) and (b), and that the coverage of the CaF_2 surface is poorer. This does not have any adverse affect on the selectivity or the reaction course because any exposed CaF_2 surface will be inert and will not promote any side reactions.

2.2.5.2. $NaBH_4$–SiO_2 An example of a type 2d process is provided by the combination of $NaBH_4$ and silica gel.[32] This has been used in an elegant method for the reduction of nitroalkenes to nitroalkanes.

This reaction is normally carried out using sodium borohydride and gives a mixture of the desired product along with variable quantities of dimer. Thus, reaction of borohydride with nitroalkene produces resonance-stabilized anion (Fig. 2.9a) that can either protonate to form nitroalkane (Fig. 2.9b) or can add to another mole of nitroalkene to give a dimer (Fig. 2.9c).

Figure 2.9 $NaBH_4$ reduction of nitroalkenes—possible reactions of intermediate (a).

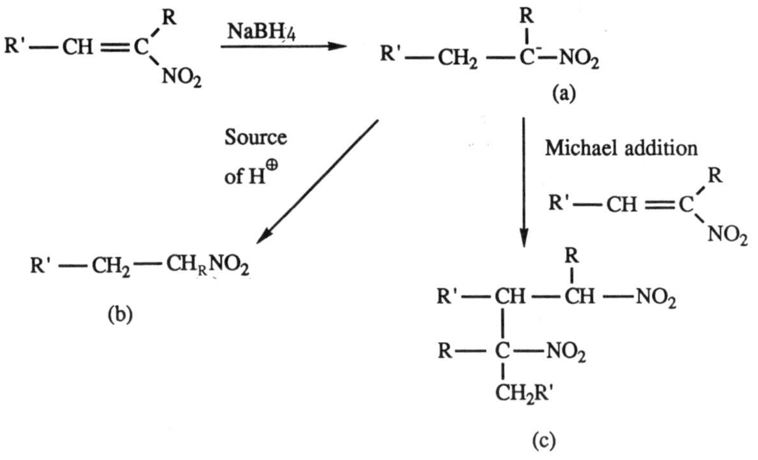

The combination of $NaBH_4$ and a source of H^+ in the same phase is difficult without first effecting the decomposition of $NaBH_4$. Sinhababu and Borchardt[32] solved this problem by having the $NaBH_4$ present in the solution phase, along with the nitroalkene, and the source of H^+ in a solid phase (i.e., silica).

A careful choice of solvent was necessary for a successful reaction. The solvent needed to be polar enough to dissolve the $NaBH_4$ and the nitroalkene/nitroalkane. For the interception of intermediate (Fig. 2.9a) before the Michael addition can occur, the solvent polarity must be as low as possible to bring about rapid adsorption of the intermediate from solution onto the silica that protonates the anion to give the desired product. The chosen solvent mixture was chloroform/isopropanol and use of this system led to excellent yields (> 90% isolated in each case).

2.2.5.3. Electrophilic Chlorination in the Presence of Silica or Zeolites. Reaction systems have been developed using either silica or zeolites as the support.[33] Various reagents, such as sulfuryl chloride,[34] *t*-butyl hypochlorite, dichlorourethane, and dichloramine-T,[33,35] have been used as reagents.

For all but the most active substrates, no chlorination occurs in the absence of support. Addition of support causes a rapid reaction to occur. The product distribution indicates that the reaction is an electrophilic chlorination rather than a radical chlorination, which is the typical reaction of sulfuryl chloride.[36] Yields of monochlorinated product are very high, with little dichlorination occurring.

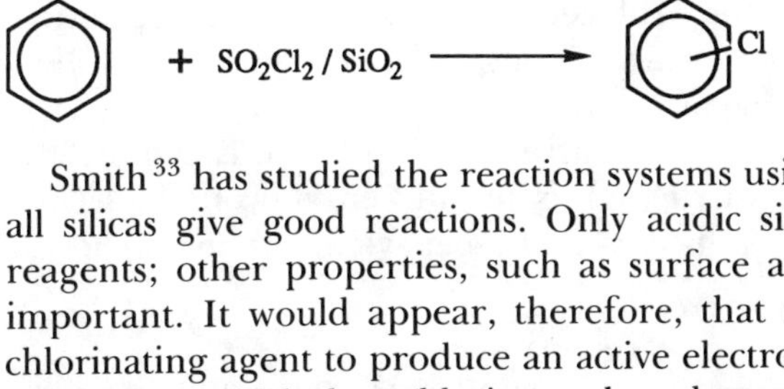

Smith[33] has studied the reaction systems using SiO_2 and found that not all silicas give good reactions. Only acidic silicas gave active supported reagents; other properties, such as surface area and porosity, were less important. It would appear, therefore, that the silica interacts with the chlorinating agent to produce an active electrophilic reagent, probably on the surface, which then chlorinates the substrate. Regioselectivities are those normally found for electrophilic chlorinations (e.g., for toluene 65% ortho: 35% para).

Because the most important property of the support appeared to be its acidity, the use of zeolites was then considered. These materials are aluminosilicates and have exchangeable cations present, typically Na^+, which can be exchanged partially for H^+. The existence of channels of molecular dimensions within the zeolite framework, coupled with the acidity of the zeolite (which is higher than that of silica) should provide the possibility of a more regioselective chlorinating system. Indeed, partially proton-exchanged Na faujasite was found to be very effective in this role, toluene giving 91% para, 9% ortho with monochlorinated products.[37]

One method of controlling reactivity is the choice of chlorinating agent. Phenol is more active than anisole and support has no effect on the chlorination of phenol when *t*-butyl hypochlorite–SiO_2 is used. Replacing *t*-butyl hypochlorite with *N*-chloro-*bis*(2-chloroethyl)amine, a less active chlorinating agent, allows the effect of the silica to become apparent. The proportion of the ortho product is higher than normal, as is the mono:dichlorination ratio.

2.2.6. *Ultrasound*

Ando and co-workers[38,39] have carried out investigations into the effect of ultrasound on the course of some reactions of KCN and alumina. These results indicate that cooperative effects can be achieved, and that the course of reaction can be altered. These results suggest that the KCN–Al_2O_3 supported reagent is being formed in situ by ultrasound.

This work concerned the displacement of halide by cyanide, promoted by alumina.[40,41] Ultrasonic irradiation of a mixture of benzyl bromide, potassium cyanide, and alumina in toluene gave a good yield of benzyl cyanide:

$$PhCH_2Br + KCN + Al_2O_3 \xrightarrow[\text{ultrasound}]{\text{toluene}} \underset{67\%}{PhCH_2CN} + KBr$$

If the alumina was omitted, the yield decreased to 27%. With no ultrasound, no benzyl cyanide was formed. Thus, both ultrasound and alumina are necessary for an efficient reaction.

In the absence of ultrasound, the course of the reaction is completely changed. Thus, mechanically stirring a mixture of benzyl bromide, KCN, and alumina without sonication gave *o*- and *p*-benzyltoluene in a yield of 87%.

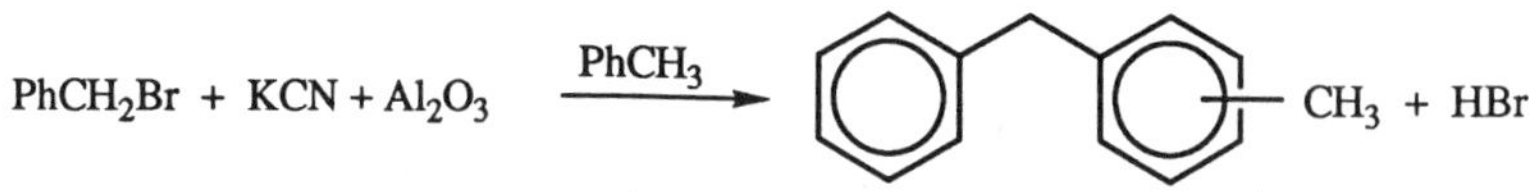

Similar results were obtained in the absence of KCN with sonication slightly enhancing the Friedel–Crafts reaction. Further variations of these

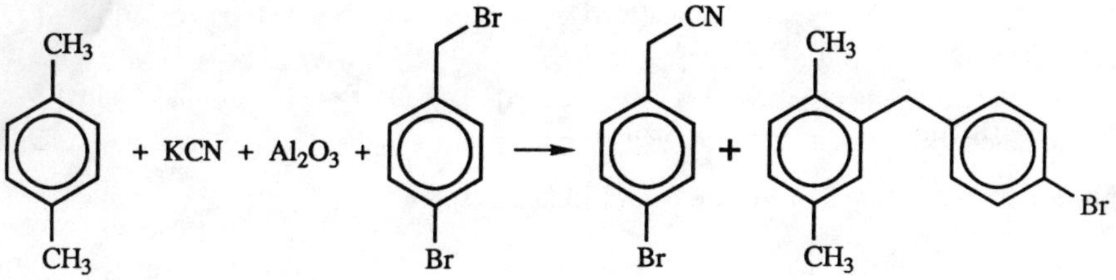

Reaction conditions		
Ultrasound (no pretreatment)	40	52
Mechanical (no pretreatment)	0	83
Ultrasound (ultrasound pretreatment)	70	8

Figure 2.10 Effect of ultrasound on the reaction path of aromatics with $KCN + Al_2O_3$.

experiments led to other important results. Three experiments were carried out where a mixture of *p*-xylene, alumina, and KCN were sonicated before addition of *p*-bromobenzyl bromide (Fig. 2.10). Reaction of benzyl bromide was then carried out using ultrasound. Thus, ultrasonic pretreatment can push the reaction path further toward bromide displacement and away from the Friedel–Crafts reaction. Interestingly, ultrasonic pretreatment, followed by reaction without ultrasound, results in no cyanation, and a much slower Friedel–Crafts reaction.

These results are consistent with the formation of KCN–alumina-supported reagent by ultrasound. This process covers the alumina surface, occluding the Friedel–Crafts sites on the alumina, and thus slowing down the formation of diarylmethanes. Ultrasound is necessary to cause the supported reagent reaction to occur, probably by conventional ultrasonic activation mechanisms, although it may have a role to play in removing products from the surface.

In conclusion, this technique seems to be capable of producing reasonably well-dispersed supported reagents, although some background support reactivity is still apparent. At the present time, however, the scope of the method has not been fully explored.

2.2.7. *Specific Syntheses*

2.2.7.1. Silica. Reactions have been carried out on silica surfaces, making use of the acidic SiOH unit to anchor reagents onto the support. The supported reagents formed are generally catalysts, or are further reacted to produce surface-bound catalysts.

1. Rhodium allyl complexes[42,43]
   A solution of yellow *tris*(allyl)rhodium was added to a toluene slurry of predried silica, adsorption of the rhodium complex onto the support was noted by the decolorization of the solution and the change in color

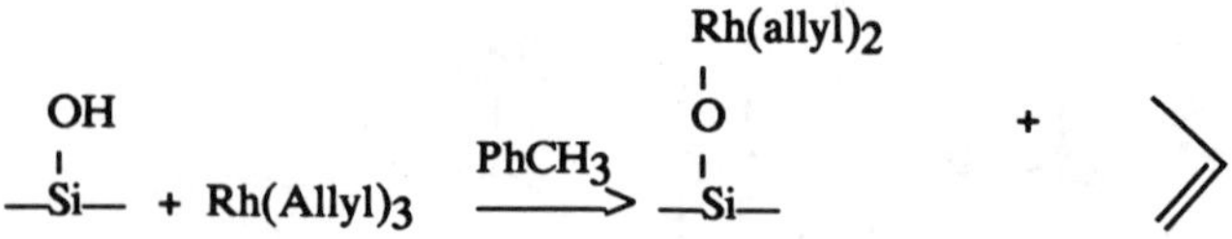

Figure 2.11 Preparation of surface bound $Rh(allyl)_2$ species.

of the silica to yellow. One equivalent of propene was also evolved. This can be represented by the scheme in Figure 2.11.

Characterization of this material has been carried out chemically and spectroscopically (see Section 2.5.1 and Chapter 3).

2. Phase transfer catalysts (PTC)
   Phosphonium salts have been immobilized successfully on silica by reaction of phosphines with surface-bound alkyl halides.[7,8] Such catalysts behave analogously to nonsurface-bound PTCs, but are easily recovered and can be reused. Various catalysts have been prepared, and their activity versus structure has been investigated. The series of reactions in Figure 2.12 were carried out after activation of SiO_2 (see section 2.1.2).

2.2.7.2. Clays. In the interlamellar region of clays there exist both SiOH units and exchangeable cations. Specific syntheses on clay supports have made use of both of these centers. An example of the second is the synthesis of Co complexes intercalated in hectorite, as reported by Davison and McWhinnie.[44] They prepared Co^{2+} exchanged clay (see Section 2.2.9) before reacting with bipyridyl and O_2 (Fig. 2.13). The work of Choudhary et al.[1,2,45] illustrates neatly the use of SiOH and SiONa groups within clays. This group has prepared a series of Pd hydrogenation catalysts by various routes. Their general methodology is the inverse of McWhinnie's (i.e., to bind a ligand to the clay, and then form a catalyst by ligand exchange of a soluble Pd complex). The success of these methods requires correct drying of the support before the reaction sequence (see Section 2.1.1). Treatment of each of these with $(PhCH_2CN)_2PdCl_2$ generated the respective Pd complexes as in Figure 2.14. All of these species are efficient hydrogenation catalysts, and their syntheses indicate the diversity of reactions possible within clay structures (Figs. 2.15 and 2.16).

2.2.8. *Substrate Adsorption*[46–65]

As an alternative to reagent adsorption, there are several papers detailing the adsorption of the substrate onto the support. The supported substrate is then reacted, often with gaseous reagents to give unusual regioselectivities. The origin of these selectivies is the ad-

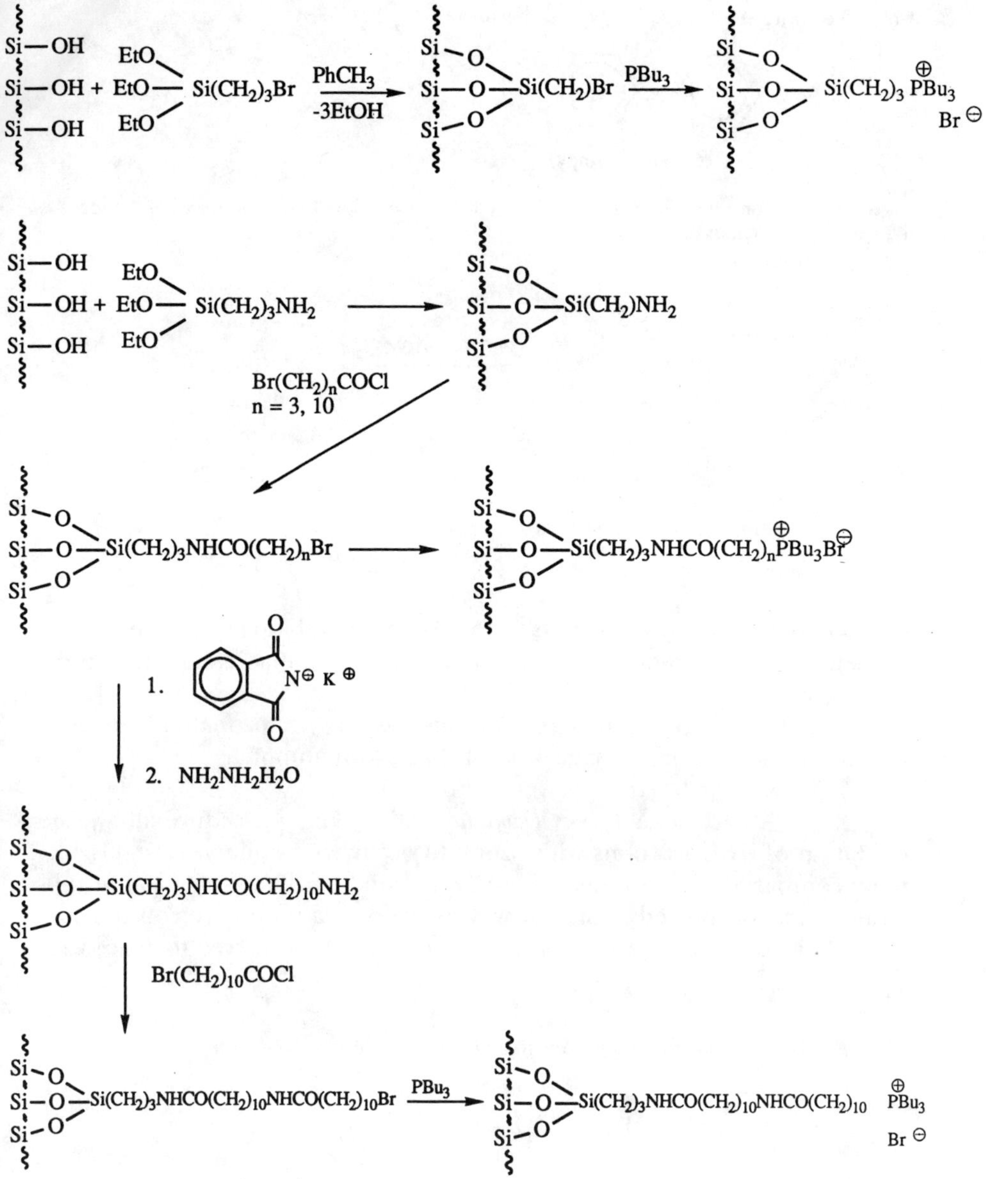

Figure 2.12 Synthesis of supported phase transfer catalysis.

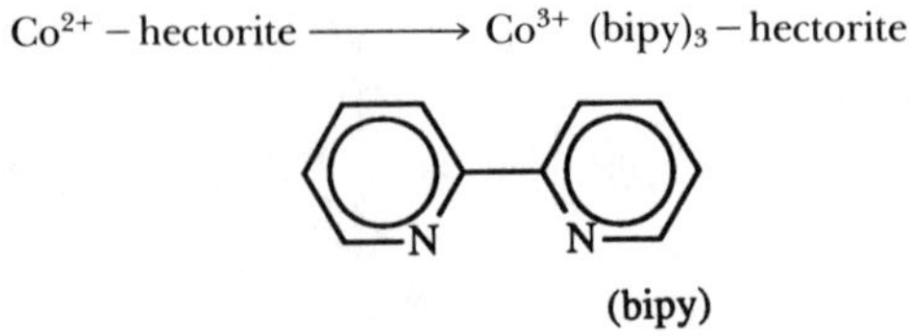

Figure 2.13 Complexation of ion exchanged cations in clays. This complex has some activity as a redox catalyst.

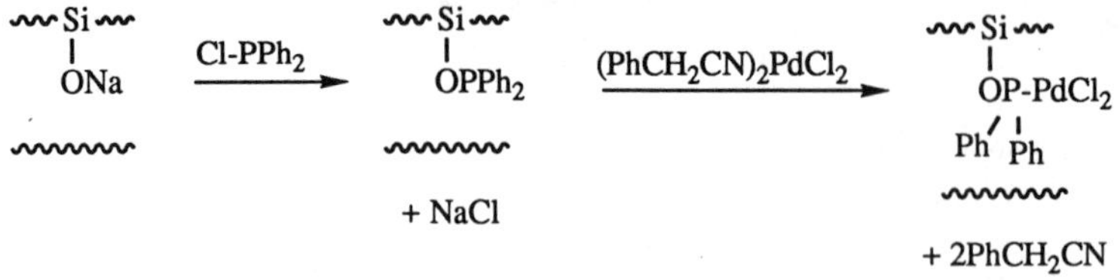

Figure 2.14 Preparation of intercalated Pd phosphine reagent.[1]

sorption of the substrate to give a monolayer such that reagents react with the surface of the monolayer, all other sites being inaccessible to the incoming reagent.

Three types of reaction system are discussed: "dry ozonation" on silica gel, chlorination, and esterification of diacids on alumina.

2.2.8.1. Dry Ozonation.[46–63] This procedure allows the oxidation of hydrocarbons with ozone to occur with enhanced regioselectivity compared with solution phase ozonation.[46,52] This is due to the substrate being supported in such a way as to form a monolayer on the surface, therefore, only certain parts of the molecule are free to react with

Figure 2.15 Reactions of SiO groups to attach ligands in interlameller space.[2]

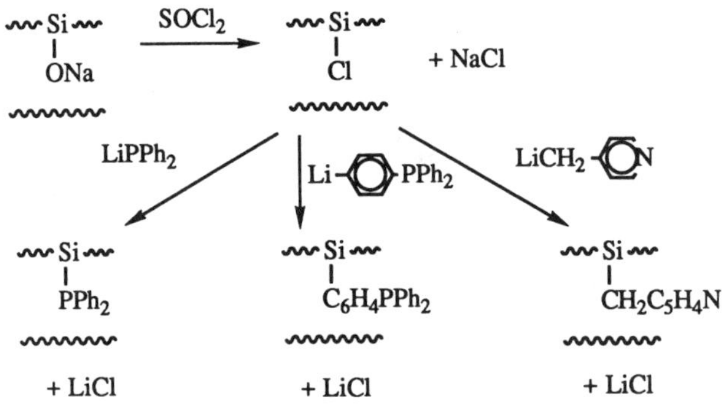

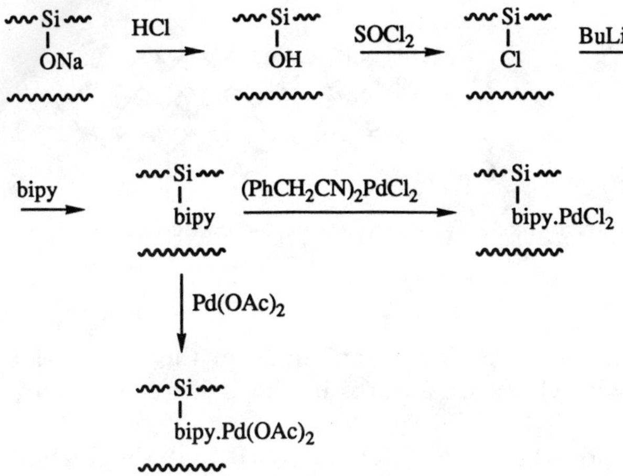

Figure 2.16 Preparation of surface bound Pd^{II} catalysts.[45]

ozone. Similar results have been achieved using crystalline steroids, where the positions at which oxidation can occur are governed by the arrangement of the molecules at the crystal surface.[64]

To achieve optimum selectivity, as close as possible to a regular adsorbed monolayer must be achieved. Too high a loading will lead to multilayer adsorption, which may be less regular and tends to reduce both yield and regioselectivity. Too low a loading will lead to isolated molecules at which all the sites may be accessible to ozone. This is illustrated in Figure 2.17. Monolayer capacities can be measured by adsorption experiments, and reaction yields and regioselectivities have been shown to be highest at or just below the measured monolayer capacities of the supports.[54]

2.2.8.2. Effect of Solvent/Polar Group. Because the surface of silica is polar, the strongest interactions between it and an organic molecule involve polar groups on the molecule. This implies that the highest

Figure 2.17 Effect of loading on regioselectivity of ozonation.

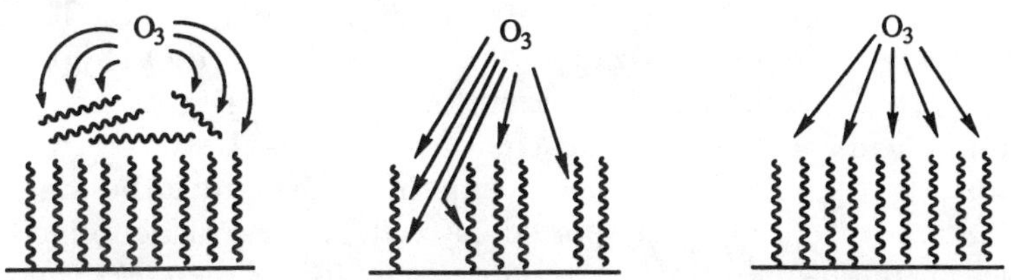

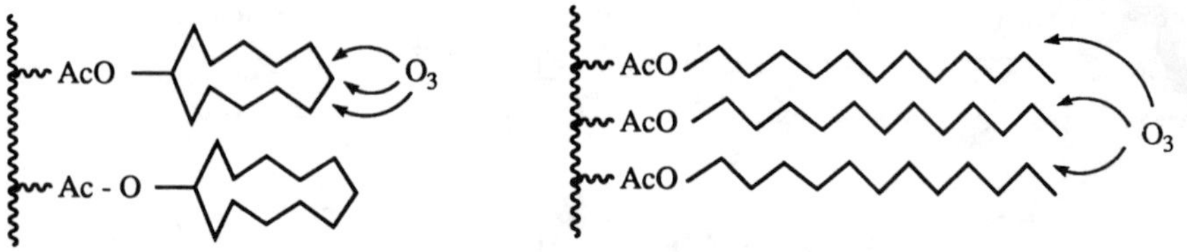

Figure 2.18 Attack of ozone on adsorbed cyclododecyl acetate and dodecylacetate.

regioselectivities are achieved at positions remote from the most polar group. Thus 3,7-dimethyloctyl acetate adsorbs by the C-1 acetate group, and reacts at C-7.

Cyclododecyl acetate adsorbs at the acetate group and the highest ozonation yields are obtained at the C-5, C-6, and C-7 sites; dodecyl acetate behaves analogously to give C-11 ketones as the product (attack at the primary C-12 position is likely to go through energetically unfavorable intermediates)[52] (Fig. 2.18). Selectivity is, therefore, decreased by solvent molecules that can overcome the surface polar group interactions. A method that strengthens the substrate–support interaction should enable reactions to be carried out in a solvent. Such a system has been developed by Beckwith and Duong.[53]

By supporting the hydrogen succinate esters of alcohols, rather than the acetates, as has been done previously, reactions could be carried out in Freon II. Hydrogen succinate esters bind more strongly to the silica support than the acetates and no solvent disruption takes place, regioselectivity being similar to the dry reactions.

2.2.8.3. Methylation of Supported Substrates. Alcohols and phenols adsorbed on silica can be methylated with diazomethane.[66] Reactions omitting the silica give no products, and other supports give much lower yields. This behavior is due to the strong polarization of the OH bond of the SiO_2.

The alcohol or phenol ($\sim$0.3 mmol/g SiO_2) is allowed to stand with silica gel for 24 h and excess diazomethane is blown through the solid. The yield of ether is high except in the case of 2,6-di-*t*-butyl phenol. This is reported to be because no polarization of the very hindered O-H bond by the SiO_2 surface is possible in this case and thus no reaction occurs.

Adsorption of acids onto supports, such as alumina, has been shown to involve chemisorbed species containing symmetric bidentate carboxylate groups.[67–71] Dicarboxylic acids, if supported correctly, can be made to bind to the surface through one acid group only, leaving the other free (Fig. 2.19).

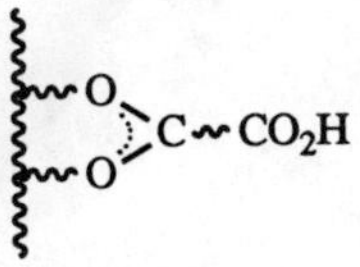

Figure 2.19 Interaction of diacid with alumina surface.

Ogawa and his group[72–74] have shown that adsorption of dicarboxylic acids on alumina can be used to give high yields of products arising from the nonbound carboxylic acid group. As before, monolayer coverage gives the best results.

Two methods of preparation have been used for the supported substrate. Both use solutions of the dicarboxylic acids in DMF (1–3%). In one, 100–300 mg of diacid in 10 ml of DMF is added to alumina and occasionally shaken for 8 h. The alumina is separated by filtration, washed with DMF to remove unbound acid and dried under vacuum for 30 min. This gives ~75% adsorption of the diacid. Alternatively,[74] after standing for 8 h the DMF can be removed under vacuum. These techniques give a monolayer of adsorbed acid, stacked perpendicularly to the surface.

Reaction with diazomethane gives the monomethyl ester in very high selectivity (e.g., for terephthalic acid ≥ 96.6%). Not surprisingly, phthalic acid does not give this behavior. The strength of the alumina–acid interaction allows the use of cyclohexane as a solvent for the esterifications, and diglyme has been used for the selective reduction of terephthalic acid to the mono-alcohol (Fig. 2.20). Other esterification reagents have been used, leading to lower selectivities. As with terephthalic acid, which is probably the most rigid acid with the carboxylate group separated by an aromatic ring, many aliphatic diacids have been successfully monoesterified by this method.[72] C-5–C-12 diacids gave the monoesters, typically in > 99% selectivity.

More recent papers have extended this concept to selective esterification of different chain length acids.[73,74] For example, dodecanedoic acid

Figure 2.20 Reactions of adsorbed terephthalic acid.

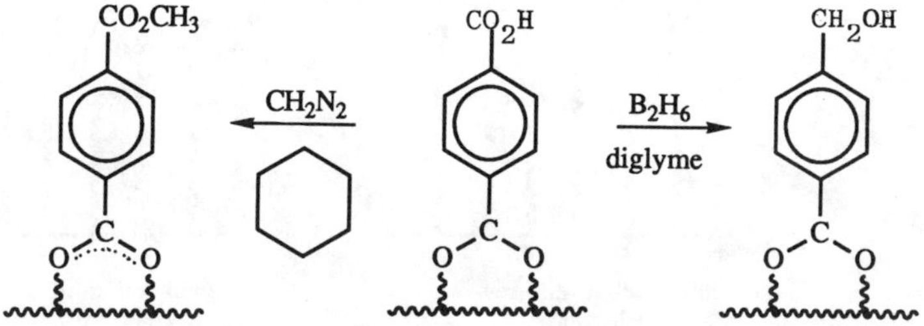

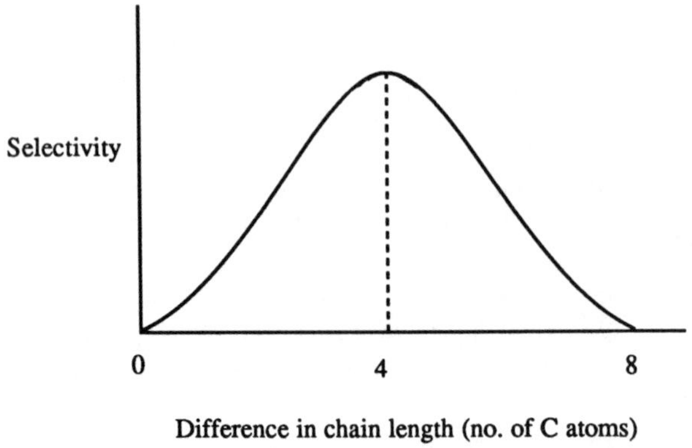

Figure 2.21 Change in selectivity with difference in chain length for mixtures of diacids.

can be monoesterified in the presence of adipic acid. This behavior is due to steric hindrance of the shorter chain diacids by the longer chain homologues. For large differences in chain length moderate selectivity is achieved, rising to an optimum as chain lengths become too similar for selective reactions to be reasonable (Fig. 2.21). The reason for this behavior can be rationalized by changes in the structure of the monolayer. Matuo et al.[75] have shown that in monolayers, interaction of pentadecanoic acid with its close homologues increases as the size of the homologue gets closer to pentadecanoic acid. For mixtures of diacids supported on alumina, this rationale would lead to the following model. For large differences in chain length, the two acids will form "self-associated" regions, their self-attraction being greater than the cross-attraction. As their chain lengths become more similar there exists a better mixed monolayer. When the chain lengths are very similar, however, all the free acid groups are available for reactions (Fig. 2.22). Selectivity can be increased with large

Figure 2.22 Effect of difference in chain length on selectivity esterification.

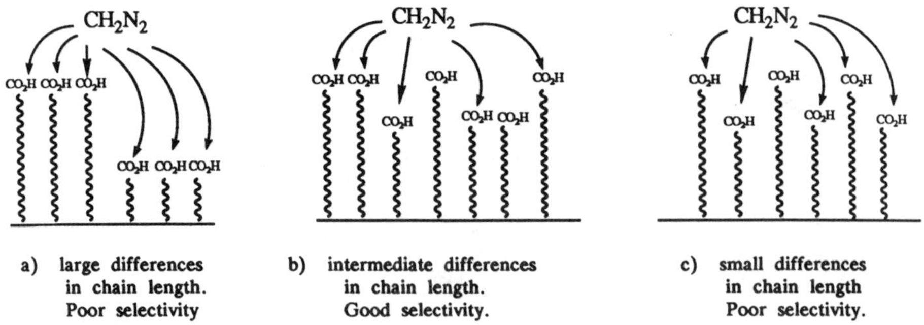

differences in chain length by increasing the amount of longer chain acid, forcing areas of short chain acids to be smaller and less accessible to diazomethane (i.e., a situation more like the one in Fig. 2.22b).

To summarize, the work described in this section emphasizes strongly the importance of correct surface coverage in the field of supported reagents. It also details how a knowledge of substrate–support interactions can be used to define optimum reaction conditions, and how successful reactions can be carried out using very delicate systems indeed.

2.2.9. *Ion Exchange*

2.2.9.1. Clays. Ion exchange is a technique that has been applied mainly to clays, materials that have significant quantities of exchangeable cations in the interlamellar regions. The exchangeable ions initially present in clays are predominantly Na^+, although in acid-treated clays significant amounts of protons may be present in place of some of the sodium ions. As has been pointed out,[76,77] H^+ exchanged clay can contain Al^{3+} ions in the interlamellar gap, caused by dealuminization of the framework by exchange with H^+. Also Al^{3+} exchanged clays may give H^+ due to hydrolytic reactions leading to polyhydroxy Al cations.[78] The extent to which ion exchange can take place varies from clay to clay. Exchangeable cations are present in clays to balance the negative charge on the framework sheets. For example, if a Si atom (formally 4^+) is replaced by an Al^{3+} or a Fe^{3+} in the framework, then this produces an overall charge of -1 in the framework, and can be counterbalanced by Na^+ or $\frac{1}{2}$ Mg^{2+}. These cations can be exchanged for others, while keeping the overall charge neutral. The number of exchangeable cations is measured by the cation-exchange capacity (CEC) normally measured in milliequivalents/100 g. Typical values for this parameter lie in the range 60–120 mEq/100 g. Normally, ion exchanges are carried out with an excess of reagent to force the equilibrium toward the product.

A “standard” procedure is given:

> 100 g of the clay is suspended in a solution containing a four-to five-fold excess of the cation to be exchanged (based on CEC), usually the chloride, although a variety of anions can be used depending on the solubility, and is stirred vigorously for several hours (3–24). The clay is separated by centrifugation and the supernatant solution discarded. The solid material is suspended in distilled water and vigorously stirred for ~ 1 h, recentrifuged, and the solid collected. Some authors repeat the entire procedure at this stage. Whether or not this is done, the final step is a repeated washing/centrifugation of the clay until no more ions are present in the washings. Drying of the solid is usually carried out to remove any water, either at atmospheric pressure (110–120°C) or in a vacuum oven at 50–80°C. Fi-

nally, the clay is ground to a fine powder. The ion-exchange clays are normally stable for several weeks.

Several points should be made regarding this procedure:

1. The solution used for the exchange is normally aqueous, although ethanol has also been used successfully for several salts.[79]
2. The concentration of the solution is usually in the range of 0.1–1 M. A wide variety of cations has been used.[77,79,80] The anion is usually nitrate, chloride, or sulfate (i.e., readily available, soluble salts are used).
3. Centrifugation is the easiest method for separation of the clay from the suspension. An additional benefit is that denser contaminants, such as silica and iron oxides, which are sometimes present, can be removed by this method.[77] If a centrifuge is not available, then filtration is an alternative. Filtration of clays from hydroxylic solvents, such as those used in ion exchange, however, can be *extremely* slow, and the repeated filtration/resuspension steps necessary can take several days compared with a few hours for centrifugation. (Interestingly, filtration of clays from dry, nonhydroxylic solvents, such as toluene, can be extremely fast!)
4. The choice of cation is an important factor, and depends on the end use of the ion-exchanged clay. The acidity of clays is reported to be due to strong polarization of water by the exchangeable cations:

$$M(H_2O)_x^{n+} \rightleftharpoons [HOM(H_2O)_{x-1}]^{(n-1)+} + H^+$$

$$\text{or } M(H_2O)_x^{n+} + H_2O \rightleftharpoons [HOM(H_2O)_{x-1}]^{(n-1)+} + H_3O^+$$

This implies that the higher the charge density of M^{n+}, the more acidic the material. Because ion-exchanged clays are normally used as solid acid catalysts, it would seem logical that highly charged species, such as Al(III), Fe(III), and Cr(III), would form the most active materials. This is often the case, although Al(III) can give variable results. This may be due to formation of complex cations, reducing the charge density.[81]

Laszlo and Mathy,[79] however, have found that in a variety of Friedel–Crafts alkylations the order of activity varies with the reaction and that divalent cations, such as Zn(II) and Cu(II), are sometimes more active than cations such as Al(III) and Fe(III). These results indicate that the chemistry of these species is more complex than that suggested by the previous equations. It should be noted that the relative reactivities of the various cations in the clays bears little relation to the corresponding Lewis acid activities ($AlCl_3 >> ZrCl_4$ but clay–Al(III) << clay Zr(IV)).

This is a good example of the effect of the clay structure changing the chemistry of the exchanged cations. Similar effects have been seen with supported clays (see Section 2.5.2.5).

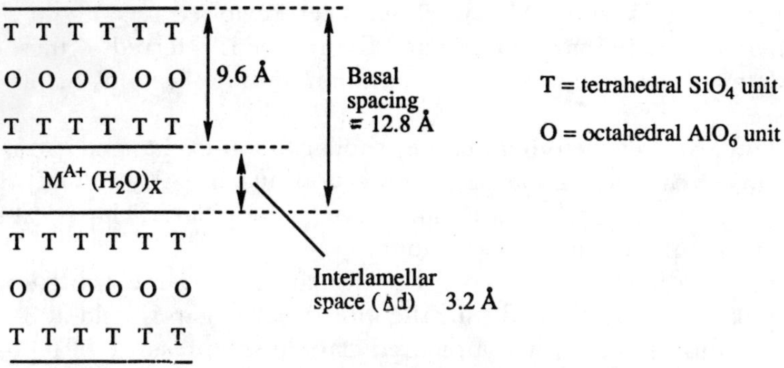

Figure 2.23 Schematic representation of montmorillonite structure.

2.2.9.2. Pillared Clays. In the ion-exchanged clays detailed previously, the naturally present cations were exchanged for simple cations such as Al^{3+}, Fe^{3+}, and Zn^{2+}. The effect on the clay's basal spacing is minimal. A typical basal spacing for a nonexchanged montmorillonite is 12.8Å. This is made up of the thickness of the clay sheet and the gap (Δd) and the next (Fig. 2.23).

The interlamellar space contains the exchangeable cations and usually the water of hydration. For a typical montmorillonite this space is 3.2Å across, given that the water of hydration is present. After ion exchange has taken place, the interlamellar space is very similar to that before ion exchange. For reactions to occur, the reactant molecules must be able to diffuse into the interlamellar space to the vicinity of the exchanged cations. This often requires that the interlamellar space be increased. This can be done by the use of a solvent that will swell the clay. The effect of solvent on the interlamellar space and the rate of reaction has been studied.[82] This work shows that certain solvents (e.g., octane) give rise to very small interlamellar spacings, and also to low reaction rates. Thus, one implication of using ion-exchanged clays is that the reaction solvent may be restricted on the basis of its swelling ability.

An alternative approach that gets around this problem is the use of pillared clays. This approach involves the ion exchange of large cluster cations into the clay. These cations hold the clay sheets apart permanently, with large interlamellar spaces in the order of 10Å. Such materials still have the acidic properties of the ion-exchanged clays mentioned. Pillared clays can also have increased thermal stabilities, being effective catalysts at temperatures that can cause interlamellar collapse in nonpillared clays.

In the last 20 years, much work has been carried out in the preparation and properties of pillared clays, and many variants, based on Al, Zr, Cr,

Ti, and Fe pillars, are known. This work has been reviewed recently in a series of papers (see References 83–96) and the reader is referred to these texts for a detailed discussion of the preparation and applications of pillared clays.

Pillared clays are prepared in a manner similar to ion-exchanged clays, the cation being exchanged is a large cluster cation such as $[Al_{13}O_4(OH)_{24}(H_2O)_{12}]^{7+}$. After exchange, washing, and drying, the pillared clay is calcined at a high temperature for a few hours.

As can be seen, the basal spacings for such materials are much higher than in nonpillared clays,[80,97] and thus the interlamellar gap is held at 8–10Å compared with 2.5–5Å for nonpillared clays in the presence of typical reaction solvents.

2.2.9.3. Silica. Whereas a large quantity of work exists detailing ion exchange in clays, little is available for other solid materials. Some work on ion-exchanged silica does, however, exist.

Silica is a high surface area material that contains, in the bulk material, tetrahedral SiO_4 networks. At the surfaces there exists a large number of Si-O-H units. These surface groups are significantly acidic, and silica has been successfully used as an acid catalyst. Ion-exchange reactions on silica make use of this acidity either directly or by salts:

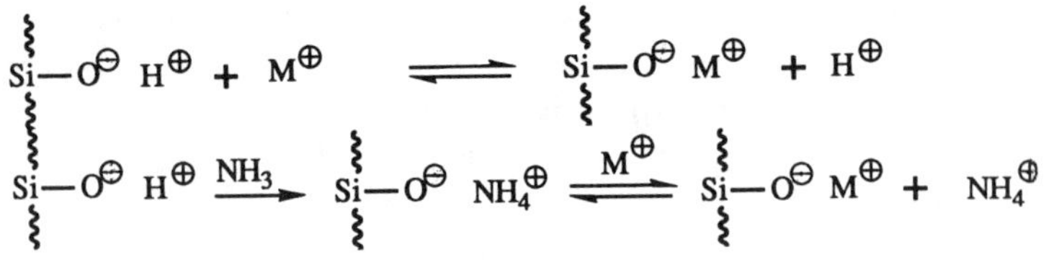

The first approach has been used to prepare a CrIII–SiO_2 polymerization catalyst.[98] This involved stirring a hot, dilute $CrCl_3$ solution in the presence of a small amount of silica (~3 g/L). Washing and drying gave the active catalyst, with Cr loading of up to 0.13 wt%

Preparation of Cu^{2+}–SiO_2 and Cu^{+}–SiO_2 has been carried out using the second method.[99] The ion exchange of $Cu(NO_3)_2$ with ammonia-treated SiO_2 was run at a lower dilution (20 g silica/L) in aqueous ammonia. The Cu concentration on the product varied from 0.11 to 5%, depending on the NH_4^+–$Cu(NH_3)_4^{2+}$ ratio in the exchange solution.

The Cu^{I}–SiO_2 product was prepared in a similar manner, using a CuCl–NH_4Cl solution in a deoxygenated ammonia solution to avoid oxidation. The exchange was carried out under N_2. To avoid the redox reaction $2Cu^+ \rightarrow Cu^{2+} + Cu$ from occurring, a piece of copper metal was also present. The final Cu content varied from 2.4 to 2.7%.

2.3. Choice of Loading

The choice of loading in the preparation of a supported reagent can be very important. In some cases the wrong loading can lead to competing reactions or to poor utilization of the reagent. Other problems can be coagulation of the supported reagent during preparation, leading to a product that is not free flowing, and therefore, is less easily handled and less efficient as problems, such as loss of activity, may be encountered.

For some common methods of preparing supported reagents, a choice of loading is available. Other methods do not allow a choice as the reagent can only occupy specific sites on the surface. For example, ion-exchanged reagents are dependent on the ion-exchange capacity of the support. It may be possible to exercise some control over the loading by using variations of the support material, but often this may lead to reagents of different properties and is not an effective method of controlling loading. For most materials the ion exchange/loading is usually the maximum possible (i.e., all the available sites are used). This normally gives the highest activity, and rarely is there an advantage in partial exchange/loading.

For other techniques, there is no maximum loading. For instance, mixing a support with a reagent can be done in any ratio. One ratio will give an optimum reagent. Changing this ratio in one direction can cause one set of problems to appear, in the other direction another set of problems may become apparent. Therefore, it is important that the correct choice of loading is made wherever possible.

This section discusses this aspect of preparing supported reagents, and highlights some of the common problems encountered that are diagnostic of incorrect loading as well as some simple techniques that can cast light on the problem.

2.3.1. Correct Loading

The correct choice of loading is often essential for a good reagent. Theoretically, this should be calculable from the surface area of the support, and knowledge (or an estimate) of the area taken up by the reagent. The surface area of the supports, however, is usually measured by adsorption of nitrogen and subsequent application of this logic. The nitrogen molecule is much smaller than almost any other species likely to be supported. For a flat surface of zero porosity, this is not a problem, as all the surface is accessible to reagents of any size. The vast majority of supports, however, are much less well defined and have varying degrees of porosity, also many of the pores may be very small. Thus, the surface area accessible to a N_2 molecule is almost certainly greater than that accessible to other species. This means that the measured surface area is a maximum, and that a varying amount of surface is inaccessible in practice. Moreover, some of the pores that are accessible to the reagent may become reduced in size to such an extent that no diffusion of substrate molecules is possible. These factors can reduce the available surface area for the preparation of the supported reagent by a considerable amount.

Because surface area measurements of the support are, therefore, not often practically helpful, more empirical tests must be employed. A useful guideline is that the vast majority of supported reagents are optimally loaded somewhere between the levels of 0.5–2 mmol/g. This seems to be true for all supports and is a useful place to start. The ultimate test for optimum activity of a supported reagent is based on reactions; however, a variety of techniques are available that can indicate the probable optimum loading. These are discussed in detail in Chapter 3. The infrared spectrum of some reagents is an easy diagnostic guide. Some supported reagents have bands for active species at different frequencies from inactive overloaded species. Several supported reagents have benefited from such an optimization procedure. Examples are sodium azide on various supports,[100] copper(I) thiocyanate on charcoal,[14] copper (I) cyanide on alumina or charcoal,[101] and potassium permanganate on silica.[102]

Two forms of adsorption are possible in the preparation of supported reagents—chemisorption and physisorption. Physisorption is a situation where the reagent is adsorbed on a surface without any chemical bonding occurring. Thus, this type of adsorption tends to be nonspecific in relation to surface chemistry. Physisorbed material in supported reagents is often present as microcrystals. This may represent the fact that the lattice energy of the crystals is greater than the energy of the support–reagent interaction, thus favoring crystal formation. The lack of any specific interaction with the surface means that the reactivity of the reagent is not greatly changed by being physisorbed (except for effects due to surface area differences, and effects caused by interaction of the substrate/intermediate/product with neighboring surface sites).

Chemisorption, on the other hand, is caused by a specific chemical interaction (i.e., bonding of some description) between the support and the reagent. In contrast to physisorption, chemisorption is normally specific to certain types of surface sites. This can often affect the reactivity of the reagent.

Relaxation of the criterion for chemisorption—chemical bonding—can lead to a simple test for chemisorbed and physisorbed material. If a soluble compound can be washed off the supported reagent by the solvent it was prepared from, washing with other solvents may cause reaction if chemisorbed species leading to solubility, it can be considered physisorbed, if not it can be considered chemisorbed. This logic, in effect, sets a lower limit to the bonding energy of a chemisorbed molecule as being equal to its free energy of solution.

The correct loading of supported reagent often corresponds to that required for monolayer adsorption. Problems may exist on either side of this optimum.

2.3.2. Underloading

A supported reagent in an underloaded state may have the following problems:

1. Greater than optimum amount of supported reagent is needed to carry out reactions. Because the loading of the supported reagent is less than optimum, then a greater mass than normal is necessary. Usually this is not a major problem, but can be easily avoided.
2. Exposed support surface. This is normally a bigger problem as supports themselves are often reactive (see Chapter 1). It is worth knowing the support chemistry, as this will normally be enough to predict the possible drawbacks and side reactions caused by support reactions, and whether they need to be avoided. (In some cases the presence of an exposed surface site adjacent to a supported molecule may actually be beneficial and lead to novel reactivity.)
   Many examples are available and a few are presented:
   Na_2S–Al_2O_3. Kamitori et al.[20] prepared a supported reagent by stirring acidic alumina in an aqueous solution of sodium sulfide. Filtration produced the supported reagent. Ando et al.[21] showed that the loading of Na_2S in this material was 0.5 mmol/g—only half of the sodium sulfide in the solution was adsorbed. Thus, the reagent was underloaded, leaving some of the alumina surface exposed. Reacting the supported reagent with aromatic aldehydes gave a Cannizzaro reaction, leading to mixtures of alcohols and (strongly adsorbed) acids. Ando et al.[21] showed that this behavior was characteristic of the support itself, and that increasing the amount of Na_2S supported (by using an evaporative technique) to > 2 mmol/g (i.e., $>$ monolayer coverage) stopped the Cannizzaro reaction.
   NaCN–Al_2O_3. The reagent NaCN–Al_2O_3 is prepared by evaporating a solution of NaCN containing alumina. This reagent is a highly efficient reagent for the nucleophilic displacement of halides to form nitriles.[40] Alumina has been shown to be the best support material for the activation of cyanide. The reaction of the NaCN–alumina reagent gives excellent yields of nitriles from alkyl bromides ($> 93\%$). The reagent as used has a 5 mmol/g loading. This is quite high and in fact only 1.8 mmol/g of cyanide is active, the remainder does not take part in the reaction and may be present as microcrystals, trapped in pores or may be unsupported. The optimal loading for the reagent is probably 1.8 mmol/g.

$$\text{R–CH}_2\text{Br} \xrightarrow[\text{PhCH}_3]{\text{NaCN/Al}_2\text{O}_3} \text{R–CH}_2\text{CN}$$

$$\text{R–CH}_2\text{Br} \xrightarrow[\text{PhCH}_3]{\text{NaCN + Al}_2\text{O}_3} \underset{\text{major}}{\text{R–CH}_2\text{OH}} + \underset{\text{alkene or ether?}}{\text{other}} + \underset{\text{minor}}{\text{R–CH}_2\text{CN}}$$

Figure 2.24 Reactions of the NaCN–Al_2O_3 system.

A high loading will, however, cover the alumina surface very efficiently[103] and, therefore, no reactions promoted by the alumina surface will occur. These reactions dominate with underloaded NaCN–alumina. Thus, when finely ground NaCN and alumina are added separately to a reaction system, very little displacement by cyanide occurs, the main reactions are those promoted by the exposed alumina (Fig. 2.24) (see also Section 2.2.6).

2.3.3. Overloading

Too high a reagent to support ratio can cause other problems. Often, this situation leads to some of the reagent being unreactive and not participating in the reaction. This may be due to some reagent crystallizing from solution, but not on the support surface. Alternatively, if maximum surface area is the main criterion, then this is achieved at the optimum loading, and increasing the loading will reduce the available surface area, probably by a combination of multilayer adsorption of the excess reagent. If the chemisorbed species has altered reactivity then this will be at best reduced, at worst lost altogether. In extreme cases of overloading the supported reagent will resemble a crystalline solid rather than a granular solid. In these cases the support is so heavily overloaded it is essentially irrelevant.

Some work has been carried out to investigate the importance of loading on the efficiency of the supported reagent. In most cases significant increases in reagent usage has been found suggesting that care taken in the choice of loading is rewarded with superior performance.

1. $KMnO_4$–silica. Two groups have studied this system. Regen and Koteel[104] prepared $KMnO_4$–SiO_2 with a loading of 2.53 mmol/g and investigated its reaction with benzyl alcohol. By studying changes in the infrared spectra of the material at various loadings, Clark et al[102] found that the optimum loading was ca. 0.51 mmol/g (i.e., approximately one fifth of

$$PhCH_2OH \xrightarrow{KMnO_4/SiO_2} PhCHO + PhCO_2H$$

Reagent loading (mmol/g)	$KMnO_4:PhCH_2OH$	Yield (%)	
		PhCHO	$PhCO_2H$
2.53	2.8:1	45	45
0.51	2.8:1	100	0
0.51	0.57:1	80	0

Figure 2.25 Effect of overloading on reagent selectivity/yield.

that used previously). At this loading the reaction with benzyl alcohol was much more efficient and selective in terms of formation of benzaldehyde (Fig. 2.25). Thus spectroscopic analysis led to a optimally loaded superior reagent that resulted in a much more selective oxidation to benzaldehyde, as well as a far better utilization of the reagent (84% compared with 21% on the basis of $MnO_4^- \rightarrow MnO_2 + 3e^-$).

2. NaN_3–various supports. The work of Onaka et al.[100] on supported sodium azide again shows that overloaded reagents are less efficient than correctly loaded reagents, using three supports—alumina, CaY zeolite, and silica. The results of the reaction of this supported reagent for the ring opening of 1-octene oxide are given in Figure 2.26. In each case, reagent loading approximately double the optimum causes the reaction yield to half—thus the utilization of the supported azide is poorer in the overloaded reagent.

Figure 2.26 Effect of overloading on yield of ring opened products.

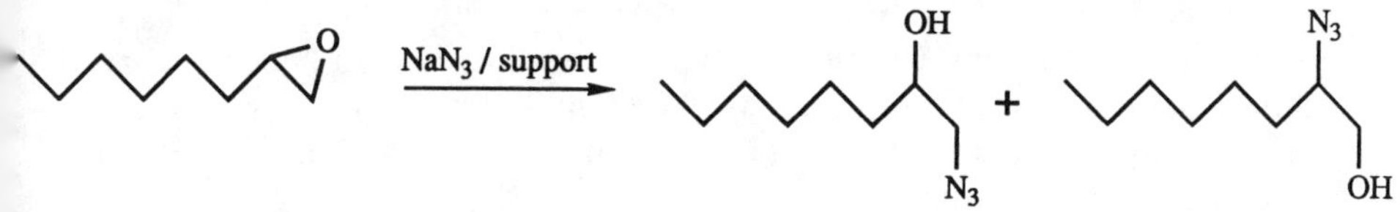

(1) (2)

Support	Loading (mmol/g)	% Yield (1 + 2)
Al_2O_3	1.6	47
	0.76	74
CaY	3.8	40
	1.9	71
SiO_2	2.3	32
	1.1	59

Studying the infrared spectra of these materials is easily done, and provides more information on the reactivity of the reagents. The supported reagent has two bands. On reaction one disappears and is thus associated with active NaN_3. The other is unchanged and is due to unreactive NaN_3, and is formed in overloaded materials. Further details are given in Section 3.2.2.

3. $Na_2Cr_2O_7$–Al_2O_3. This example shows that overloading of supported reagents can cause activity loss because of the coverage of the active chemisorbed species. Sodium dichromate supported on alumina by an evaporative route is a catalyst for the oxidation of aromatic hydrocarbons such as diphenylmethane.[22] Loadings of up to ~3 mmol/g gave a catalytic activity that varied very little. Water washing of these supported reagents gave a chemisorbed-supported reagent with a loading of ~0.075 mmol/g. This material gave a slightly higher activity in terms of substrate oxidation per gram of supported reagent, a behavior that suggests that the catalytically active species is the chemisorbed material (a conclusion that is consistent with the noncatalytic behavior of unsupported sodium dichromate). Interestingly, the chemisorbed material is accessible at loadings up to 3 mmol/g. Increasing the loading to 10 mmol/g (severely overloaded) causes the *catalytic* activity to be completely lost. This result shows that excessive overloading cannot only give essentially unreactive material but can inhibit the reactivity of the active species by rendering it inaccessible to the reactants.

2.4. *Choice of Solvent*

In general, solvents for supported reagent preparations are chosen from simple considerations. Thus, reagent solubility and compatibility with the reagent and support is necessary. Volatility is normally useful, certainly in evaporative processes, although it is not as important in precipitation reactions, where excess solvent can be filtered and washed from the supported reagent. For lamellar supports, such as clays, it is also important that the solvent open up the layers to allow entry of the reagent into the interlamellar gap.

Few studies have been carried out on the detailed role of solvents in the preparation of supported reagents, and it may be that, in time, major solvent effects will be found to be involved in the preparation of some supported reagents. Indeed, one such example is known already.

Supporting $AlCl_3$ on a variety of supports has been shown to produce a very highly acidic solid.[19] Unlike most of the known $AlCl_3$–inorganic oxide materials[105–107] this material does not evolve $AlCl_3$ with time. The evolution of HCl in equimolar amounts indicates that a specific surface-bound species is formed. The solvent involved in the preparation of this supported reagent plays a very important part in this process. Carbon

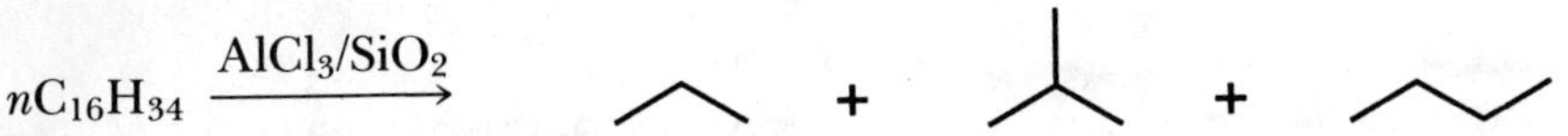

Preparation solvent	Relative amounts propane	i-butane	*n*-butane
CCl_4	1	17	1.3×10^{-3}
$CHCl_3$	1.4×10^{-6}	9.8×10^{-3}	9.8×10^{-3}
CH_2Cl_2	4.3×10^{-6}	1.7×10^{-4}	7.9×10^{-5}
$ClCH_2CH_2Cl$	5.7×10^{-6}	0	0
C_6H_6	2.4×10^{-6}	0	0

Figure 2.27 Relative activities of $AlCl_3$–SiO_2 prepared from different solvents.

tetrachloride as solvent provides the most active acid catalyst with all other solvents being inferior. This behavior is exemplified by the relative cracking abilities of the supported $AlCl_3$ reagents shown in Figure 2.27. Thus, it is obvious that the solvent plays a dramatic role in the preparation of this particular reagent; however, how general this behavior is in other systems is not yet known.

2.5. Supported Reagent Posttreatment

Posttreatment of supported reagents is often a necessary step in their preparation. Most posttreatments can be divided initially into two types: further reactions or thermal posttreatment.

2.5.1. Further Reactions

Further reactions have normally been carried out on chemisorbed species, usually prepared by adsorption from solution, although the technique could be equally applied to other reagents. This technique usefully extends the range of reagents that can be prepared, and can also be used as a method of exploring the nature and reactivity of the supported species. The work of Choudhary[1,2,4,5] provides a good example of posttreatment, as does the work of Tundo and Venturelli.[7,8] Both are discussed in detail in Section 2.2.7. In all these papers, an initial surface-specific reaction was used to anchor a reagent to the support. After this first step to prepare the initial supported reagent, subsequent reactions were carried out to produce complex supported reagents not easily accessible by more direct methods of supported reagent preparations (Fig. 2.28).

Further reactions of the silica-bound rhodium allyl complex,

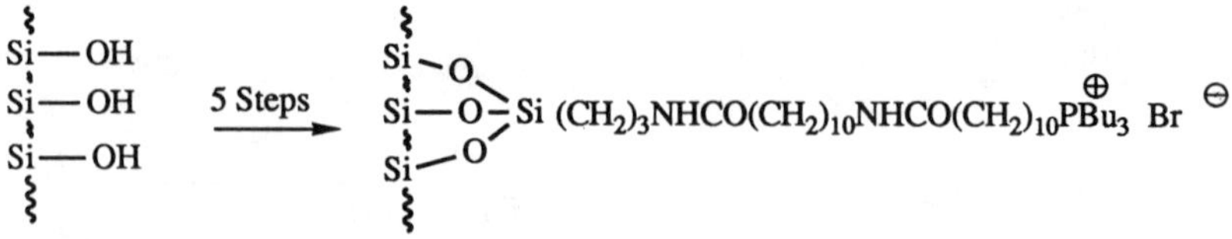

Figure 2.28 Preparation of silica-bound phase transfer catalyst.

mentioned[18,42] in Section 2.2.7, have provided a basis for the characterization of this supported reagent, and have been valuable in describing its reactivity.

The reaction of *tris*-allylrhodium with silica can be described by the equation Fig. 2.29.
It was found[42] that 1.1 mole of propene was evolved in this reaction (i.e., 10% more than the amount required by the equation in Fig. 2.29).

Posttreatment by the reaction of the supported reagent with CO indicated that in addition to the *bis*(allyl) rhodium surface-bound species, another rhodium species was present as absorption of CO gave a rhodium–CO complex without affecting the allyl species. Evidence suggests some form of Rh(III) surface-bound species (e.g,, A that reacts with CO to give the dicarbonyl shown in Fig. 2.30). Reaction of the initial supported reagent with hydrogen did not lead to supported rhodium hydrides, but to well-dispersed rhodium metal clusters that were catalytically active in hydrogenation reactions.

These examples serve to illustrate two points:

1. Post-treatment of supported reagents by further reactions can lead to many new supported reagents either inaccessible by a direct route or different to the equivalent from a direct route.

Figure 2.29 Formation of silica-bound bisallyl rhodium.

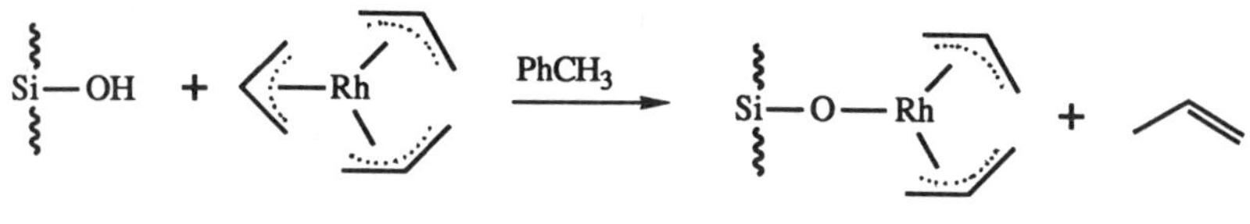

Figure 2.30 Posttreatment of surface-bound Rh species.

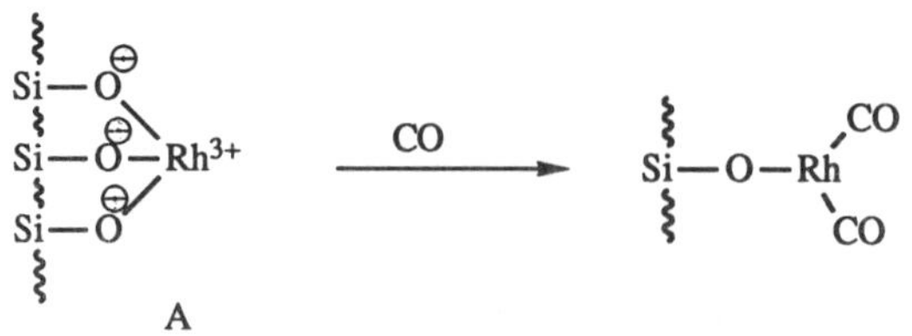

A

2. A lot of information about the nature of supported reagents can be obtained by further reactions. This information can be of use both in the characterization of the supported reagent and in a definition of its activity. This information can be important for optimization and use of the supported reagent.

2.5.2. *Thermal posttreatment*

A much more common method of posttreatment of supported reagents is, broadly speaking, thermal "activation." Because this form of posttreatment is the most convenient way of controlling the water content of a supported reagent, the effect of water is discussed here. The scope of thermal posttreatment is wider, and other effects, not related to drying, can be produced by heating supported reagents.

2.5.2.1. Role of Water. The majority of supported reagents are posttreated by drying. This is normally done by heating, with or without vacuum, to 100°–120°C. In other cases, lower temperatures are used in conjunction with either a vacuum or a dessicator system. These treatments will certainly reduce the amount of water present, but the residual water content is generally not known. There is a considerable amount of evidence available suggesting that this parameter is, in fact, very important to the activity of the supported reagent. Water, especially when it is polarized by medium to strong hydrogen bonds, can be quite reactive and may interfere by taking part in competing reactions such as hydrolysis. Thus, reducing the level of water is often a sensible step.

2.5.2.2. The Optimum Amount of Water—Water as an Activator. It appears that for many supported reagents, too great a loss of water will result in a considerable loss in activity, therefore, an optimum amount of water is necessary for activity.

For example, Regen and Mehrotra[23] found that the combination of KOH and Al_2O_3 was effective in the hydrolysis of esters. The best reagent was prepared by crushing dried (110°C/0.1 mmHg) Al_2O_3 with KOH and using the product immediately. This gave an 85% yield of hydrolysis product:

$$PhCO_2Me \xrightarrow{KOH/Al_2O_3} PhCO_2^-K^+ + MeOH$$

A reagent containing more water than this gave a 40% yield (showing that rather than promoting the hydrolysis, extra water deactivates the supported reagent). A combination of 2 g KOH, 1.5 ml H_2O, and 2 g Al_2O_3 dried at 60°C/0.1 mmHg gave a 54% yield of product. Drying the reagent at 110°C resulted in no hydrolysis reaction.

These examples show that the optimum reagent can be deactivated either by the addition or the removal of small amounts of water.

Clark et al.[31] noted a similar behavior in the $KF–CaF_2$ supported reagent. Moderate drying resulted in an active reagent for the fluorination of alkyl and acyl halides.

$$\mathrm{RBr} \xrightarrow[\text{MeCN or Sulpholane}]{\mathrm{KF/CaF_2}} \mathrm{RF} \qquad 71–92\%$$

$R = PhCH_2, C_{12}H_{25}, PhCO, MeCO$

Less extensive drying resulted in an inferior reagent, as did more severe drying conditions.

Onaka et al.[100] have undertaken a detailed study of the NaN_3–support system. They prepared supported reagents by combining NaN_3 and Al_2O_3, CaY zeolite, or SiO_2 and dried each reagent under varying conditions. The amount of residual water was measured by TGA from 20 to 220°C. This allowed a more quantitative analysis of the role of water. For optimally loaded reagents the optimum amount of water varied with support from ~20 to ~30% based on total supported reagent weight. The reaction used to monitor reactivity was the nucleophilic ring opening of epoxides (fig. 2.31).

Figure 2.31 Effect of water on reactivity of supported NaN_3.

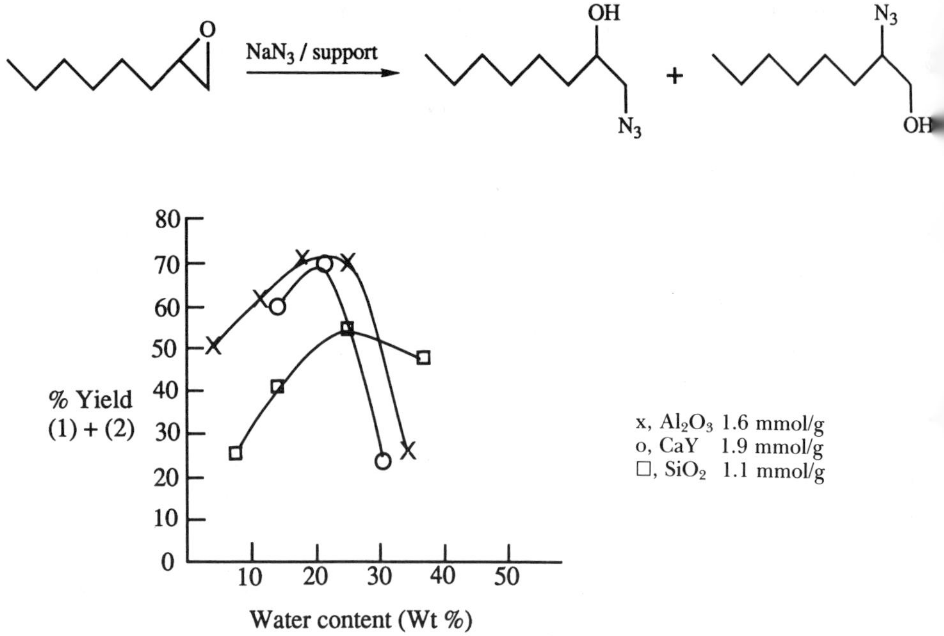

 $\xrightarrow{NaN_3/SiO_2}$ Hydroxy azide + diol

Amount of water (wt %)	% hydroxy azide	% diol
42	49	0
29	59	2.8
16	40	20
11	25	40

Figure 2.32 Reaction selectivity as a function of water content.

It can be seen that water content plays a significant part in the activity of all of the supported reagents studied. In these cases there is a well-defined optimum amount of water, with substantial decreases in activity on each side of the optimum. The competing reaction, ring opening of the epoxide by water to give the diol, is most prevalent at *low* water content. This is presumably because, as the amount of water decreases the polarization of the remaining water increases, making the water more reactive. This trend is most apparent for the NaN_3–SiO_2 reagent (Fig. 2.32). A possible explanation for the role of water in supported reagents has been proposed by Ando et al.[108]

A study of the in situ $KOAc/Al_2O_3$ reagent in the following reaction was carried out. The following results were obtained (Fig. 2.33). Again, two important points are apparent. First, the yield of acetate product increases from 1.8 to 66% as the water content is increased to the optimum value of 0.4 : 1. This yield drops to 2.3% by further increasing the amount of water. No other reaction takes place; the extra water genuinely deactivates the system as a whole.

The second point is that the amount of alcohol produced by hydrolysis of either the acetate or the bromide is highest at low amounts of water.

Figure 2.33 Reactivity/selectivity of $KOAc$–Al_2O_3 at varying water content.

$$C_8H_{17}Br + KOAc/Al_2O_3 \xrightarrow{PhCH_3} C_8H_{17}OAc + KBr$$

Molar ratio $H_2O:KOAc$	% Yield Acetate	% Yield Alcohol	Dioctyl ether
0	1.8	0.1	16
0.1	26	6.2	2.5
0.23	49	5.1	1.1
0.4	66	2.8	0.6
0.7	13	1.7	0.3
1.0	2.3	1.9	0.1

Salt	Optimum amount of water (molar ratio of H_2O : salt)	Lattice energy ($kJ\ mol^{-1}$)
CsOAc	0.23	682
KOAc	0.4	686
NaOAc	~0.6	761
LiOAc	~1	881

Figure 2.34 Lattice energy of salt versus optimum amount of water for activity.

Again, this illustrates that the reactivity of water can increase at low levels, probably due to it being more polarized, and hence more nucleophilic. Thus, the optimum amount of water for this supported reagent is a molar ratio of H_2O : KOAc of 0.4.

Other cations were then studied. The work showed that reactivity depends on the cation. The most reactive salt, CsOAc, required the least amount of water, and the optimum water content increased from CsOAc to KOAc, to NaOAc, and was highest for LiOAc. The optimum was most sharply defined for the most active salt and was broadest for the least active.

Comparison of this data with the lattice energies for the salts is very informative (Fig. 2.34). These results imply that a primary role of water may be to disrupt the lattice by hydration to produce the active species in the reaction. Such a mechanism of action has been proposed by Weber and Gokel[109] as being important in solid–liquid phase transfer catalysis and by Tanaka and Koyagani[110] in their work on heterogeneous preparation of aroyl cyanides. For water to play such a role, the reagent must exist as supported microcrystals rather than as a monolayer.

2.5.2.3. Water as a Reagent. In other cases, water may be necessary because it is a reagent itself, or it may react with one or more components of the supported reagent, producing the actual active reagent.

Water as a reagent, supported on clays, has been studied by Thomas et al.[78] Thus, wet Al^{3+}-exchanged montmorillonite or Al^{3+}-exchanged laponite hydrates 2-methylpropene to give *t*-butanol. Replacing the water with methanol allows the production of methyl *t*-butyl ether. Dehydration of the clay changes the course of the reaction to produce oligomers of 2-methylpropene (Fig. 2.35). In this example the water behaves as a reagent, thus its presence is quite important. The amount of water used was 12% wt/wt—about the equilibrium amount often found in untreated clays. No variations in amount of water were studied.

2.5.2.4. Water as a Coreagent—Possible Role in the Preparation of the Active Species. Hojo and Masuda[111] have reported a case of water changing the course of a reaction. When methyl aryl sulfides are

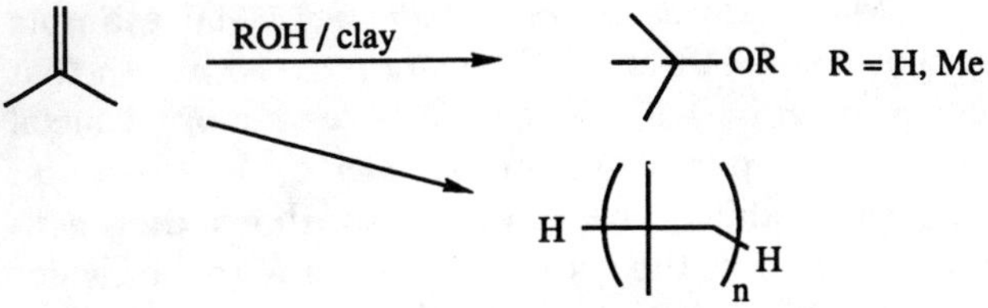

Figure 2.35 Effect of water on reactions of clay with 2-methylpropene.

reacted with SO_2Cl_2–dry SiO_2 in dichloromethane at room temperature, chloromethyl aryl sulfides are produced. This is a typical reaction of sulfuryl chloride. In contrast, using wet silica in place of dry silica, the same reaction gives methyl aryl sulfoxides, often in 100% yields:

$$\mathrm{ArSMe} \xrightarrow[\mathrm{CH_2Cl_2}]{\mathrm{SO_2Cl_2\text{–}dry\ SiO_2}} \mathrm{ArSCH_2Cl}$$

$$\mathrm{ArSMe} \xrightarrow[\mathrm{CH_2Cl_2}]{\mathrm{SO_2Cl_2\text{–}wet\ SiO_2}} \mathrm{Ar\overset{O}{\overset{\|}{S}}CH_3}$$

It is likely that the water reacts with the sulfuryl chloride to generate the active species in this system. The nature of this species is not known but may be due to hydrolysis of the sulfuryl chloride to, for example, chlorosulfuric acid or the production of ·OH radicals by homolysis of sulfuryl chloride.

In conclusion, the presence of water can be central to the activity of supported reagents. Often a specific amount is necessary for optimum activity, requiring a specific, controlled drying (or wetting) procedure. Cases are also known where the water plays a more direct part—either as a reagent itself or as a precursor to an active species.

Figure 2.36 Activation temperature versus activity for clay supported $CuCl_2$.

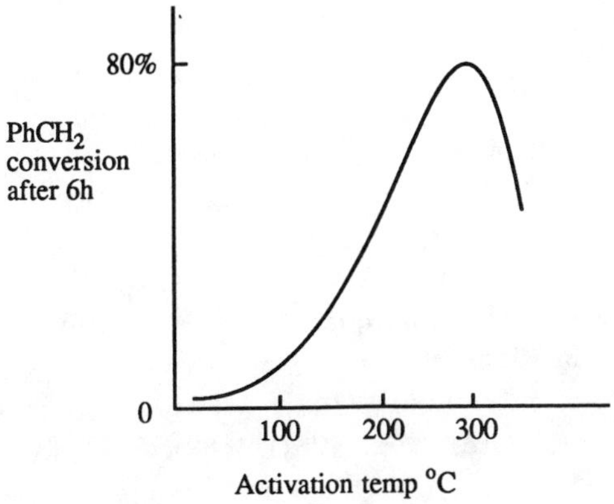

2.5.2.5. More Extreme Thermal Activation. An example of a supported reagent that requires thermal activation greater than that required to remove water is given by K10 montmorillonite-supported metal halides.[11] These materials are prepared by an evaporative route from an alcoholic solution of the metal halide. The resultant powder is then activated at atmospheric pressure to give the active material, a highly efficient Friedel–Crafts catalyst. For claycuc ($CuCl_2$–K10) the rate of reaction of benzyl chloride and benzene to give diphenylmethane is highly dependant on activation temperature, the optimum being 280°C (Fig. 2.36).

REFERENCES

1. B. M. Choudary, J. Chem. Soc., Chem. Commun., (1985), 931.
2. K. R. Kumar, B. M. Choudary, Z. Jamil and G. Thayagarajan, J. Chem. Soc., Chem. Commun., (1986), 130.
3. A. Kryswicki and M. Marczewski, J. Chem. Soc., Faraday 1, (1984), **76,** 1311.
4. G. Suzukamo, M. Fukao and M. Minobe, Chem. Lett., (1987), 585.
5. I. Mochida, T. Miyazaki, T. Takagi and H. Fujitsu, Chem. Lett., (1985), 833.
6. G. H. Posner and G. M. Gurria, J. Org. Chem., (1976), **41,** 578.
7. P. Tundo and P. Venturelli, J. Am. Chem. Soc., (1979), **101,** 6606.
8. P. Tundo, J. Chem. Soc., Chem. Commun., (1977), 641.
9. J. S. Fritz and J. N. King, Analyt. Chem., (1976), **48,** 570.
10. A. Cornélis and P. Laszlo, Synthesis, (1985), 909.
11. J. H. Clark, A. P. Kybett, D. J. Macquarrie, S. J. Barlow and P. Landon, J. Chem. Soc., Chem. Commun., (1989), 1353.
12. M. Fetizon and M. Golfier, C. R. Acad. Sc. Paris, (1968) t. 267 900.
13. J. H. Clark and C. W. Jones, J. Chem. Soc., Chem. Commun., (1987), 1409.
14. J. H. Clark, C. W. Jones, C. V. A. Duke and J. M. Miller, J. Chem. Soc., Chem. Commun., (1989), 81.
15. T. Ando, D. G. Cork, M. Fujita, T. Kimura and T. Tatsuno, Chem. Lett., (1988), 1877.
16. J. San Filippo and C-I. Chern, J. Org. Chem., (1977), **42,** 2182.
17. P. A. Risbood and D. M. Ruthven, J. Org. Chem., (1979) **44,** 3969.
18. M. D. Ward, T. V. Harris and J. Schwarz, J. Chem. Soc., Chem. Commun., (1980), 357.
19. R. S. Drago and E. E. Getty, J. Am. Chem. Soc., (1986), **110,** 3311.
20. Y. Kamitori, M. Hojo, R. Masuda and M. Yamamoto, Chem. Lett., (1985), 253.
21. T. Ando, J. H. Clark, D. G. Cork and T. Kimura, Chem. Lett., (1987), 1315.
22. J. H. Clark, A. P. Kybett, P. Landon, D. J. Macquarrie and K. Martin, J. Chem. Soc., Chem. Commun., (1989) 1355.
23. S. L. Regen and A. K. Mehrotra, Synth. Commun., (1981), **11,** 413.
24. N. A. Noureldin and D. G. Lee, Tetrahedron Lett., (1981), 4889.

25. F. Texier-Boullet and M. Lequitte, Tetrahedron Lett., (1986), 3515.
26. J. Ichihara, T. Matsuo, T. Hanafusa and T. Ando, J. Chem. Soc., Chem. Commun., (1986), 793.
27. A. J. Vogel, "Elementary Practical Organic Chemistry," Longmans, Green and Co., Toronto, (1957), p. 419–502.
28. C. V. A. Duke, J. M. Miller, J. H. Clark and A. P. Kybett, J. Mol. Catal., (1990), **62,** 233.
29. J. H. Clark, E. M. Goodman, D. K. Smith, S. J. Brown and J. M. Miller, J. Chem. Soc., Chem. Commun., (1986), 657.
30. N. Ishikawa, T. Kitazume, T. Yamazaki, Y. Mochida and T. Tatsuno, Chem. Lett. (1981), 761.
31. J. H. Clark, A. J. Hyde and D. K. Smith, J. Chem. Soc., Chem. Commun., (1986), 791.
32. A. K. Sinhababu and R. T. Borchardt, Tetrahedron Lett., (1983), 227.
33. K. Smith, Bull. Soc. Chim. France, (1989), 272.
34. M. Hojo and R. Masuda, Synth. Commun., (1975), **5,** 169.
35. K. Smith and M. Butters, Synthesis, (1985), 1157.
36. M. S. Kharasch and H. C. Brown, J. Am. Chem. Soc., (1939), **61,** 2142, 3421.
37. K. Smith, M. Butters and B Nay, Synthesis, (1985), 1157.
38. T. Ando, T. Kawate, J. Ichihara and T. Hanafusa, Chem. Lett., (1984), 725.
39. T. Ando, S. Sumi, T. Kawate, J. Ichihara and T. Hanafusa, J. Chem. Soc., Chem. Commun., (1984), 439.
40. S. L. Regen, S. Quici and S-J Liaw, J. Org. Chem., (1979), **44,** 2029.
41. G. Bram, T. Fillibeen-Khan and N. Reranghty, Synth. Commun., (1980), **10,** 279.
42. H. C. Foley, S. J. DeCario, K. D. Tau, K. J. Chao, J. H. Onuferko, C. Dybowski and B. C. Gates, J. Am. Chem. Soc., (1983), **105,** 3074.
43. M. D. Ward, T. V. Harris and J. Schwartz, J. Chem. Soc., Chem. Commun., (1980), 357.
44. N. Davison and W. R. McWhinnie, Inorg. Chim. Acta., (1987), **131,** 9.
45. B. M. Choudary and P. Barathi, J. Chem. Soc., Chem. Commun., (1987), 1505.
46. A. L. J. Beckwith, C. L. Bodkin and T. Duong, Aust. J. Chem., (1977), **30,** 2177.
47. Z. Cohen, E. Keinan, Y. Mazur and T. H. Varkony, J. Org. Chem., (1975), **40,** 2141.
48. E. Akiyama, M. Tada, T. Tsuyuki and T. Takahashi, Chem. Lett., (1978), 305.
49. E. Akiyama, M. Tada, T. Tsuyuki and T. Takahashi, Bull. Soc. Chim. Jpn., (1979), 164.
50. E. Suokas and T. Hase, Acta. Chem. Scand., B, (1978), **32,** 623.
51. J. P. Desvergne and H. Bonas-Laurent, J. Catal., (1977), **51,** 126.
52. A. L. J. Beckwith and T. Duong, J. Chem. Soc., Chem. Commun., (1978), 413.

53. A. L. J. Beckwith and T. Duong, J. Chem. Soc., Chem. Commun., (1979), 690.
54. A. L. J. Beckwith, C. L. Bodkin and T. Duong, Chem. Lett., (1977), 425.
55. E. Proksch and A. de Meijere, Angew. Chem. Int. Ed. (Eng), (1976), **15,** 761.
56. I. E. De Besten and T. H. Kinstle, J. Am. Chem. Soc., (1980), **102,** 5968.
57. D. Tal, E. Keinan and Y. Mazur. J. Am. Chem. Soc., (1979), **101,** 502.
58. E. Trifillieff, L. Bang and L. Ourisson, Tetrahedron Lett., (1977), 2991.
59. J. P. Desvergne and H. Bouas-Laurent, E. V. Blackburn and R. Lapouyede, Tetrahedron Lett., (1974), 947.
60. H. Klein and A. Steinmetz, Tetrahedron Lett., (1975), 4249.
61. E. Prokesch and A. de Meijese, Tetrahedron Lett., (1976), 4851.
62. T. Preuss, E. Proksch and A. de Meijere, Tetrahedron Lett., (1978), 833.
63. E. Keinan and Y. Mazur, J. Org. Chem., (1977), **42,** 844.
64. A. L. J. Beckwith, Proc. Chem. Soc., (1958), 194.
65. N. C. Deno, R. Fishbein and C. Pearson. J. Am. Chem. Soc., (1970), **92,** 1451.
66. H. Ogawa, T. Hagiwara, T. Chihara, S. Teratani and K. Taya, Bull. Chem. Soc., Jpn., (1987), **60,** 627.
67. S. Hayashi, T. Takenaka and R. Cotoh, Nippon Kagaka Zasshi, (1967), **88,** 133.
68. L. H. Little, "Infrared Spectra of Adsorbed Species," Academic Press, London, (1986).
69. K. Hirota, K. Fueke, Y. Nakai and K. Shindo, Bull. Chem. Soc. Jpn., (1958), **31,** 780.
70. Y. Noto, K. Fukuda, T. Onishi and K. Tamaru, Trans. Farad. Soc., (1967), **43,** 2300.
71. J. T. Hall and P. K. Hansima, Surf. Sci. (1978), **77,** 61.
72. H. Ogawa, T. Chihara and K. Taya, J. Am. Chem. Soc., (1985), **107,** 1365.
73. H. Ogawa, T. Chihara, S. Teratani and K. Taya, J. Chem. Soc., Chem. Commun., (1986), 1337.
74. H. Ogawa, N. Hirage, T. Chihara, S. Teratani and K. Taya, Bull. Chem. Soc. Jpn., (1988), **61,** 2383.
75. H. Matuo, K. Motomura and R. Matuusa, Bull. Chem. Soc. Jap., (1981), **54,** 2205.
76. J. P. Beckman and H. Laudelot, Kolloid 2, (1961), **178,** 99.
77. J. A. Ballantine, J. H. Purnell, M. Rayanakorn, K. J. Williams and J. M. Thomas, J. Mol. Cat. (1985), **30,** 373.
78. J. A. Ballantine, W. Jones, J. H. Purnell, D. T. B. Tennakoon and J. M. Thomas, Chem. Lett., (1985), 763.
79. P. Laszlo and A. Mathy, Helv. Chim. Acta, (1987), **70,** 577.
80. H. Sakurai, K. Urabe and Y. Izumi, J. Chem. Soc. Chem. Commun., (1988) 1519.
81. J. M. Adams, T. V. Clapp and D. E. Clement, Clay Miner, (1983), **18,** 411.
82. R. W. McCabe, J. M. Adams and K. Martin, J. Chem. Res.(S), (1985), 356.

83. D. E. W. Vaughan, Catal. Today, (1987), **2,** 187.
84. C. J. B. Mott, Catal. Today, (1987), **2,** 199.
85. S. L. Jones, Catal. Today, (1987), **2,** 209.
86. J. Sterte, Catal. Today, (1987), **2,** 219.
87. G. J. J. Bartley, Catal. Today, (1987), **2,** 233.
88. M. S. Tzou and T. J. Pinnavaia, Catal. Today, (1987), **2,** 243.
89. S. Yamanaka and M. Hattori, Catal. Today, (1987), **2,** 261.
90. C. I. Warburton, Catal. Today, (1987), **2,** 271.
91. J. J. Fripiat, Catal. Today, (1987), **2,** 281.
92. E. Kikuchi and T. Matsuda, Catal. Today, (1987), **2,** 297.
93. M. L. Occelli and R. J. Rennard, Catal. Today, (1987), **2,** 309.
94. H. Ming-Yuan, L. Zhonghui and M. Enze, Catal. Today, (1987), **2,** 321.
95. M. L. Occelli, Catal. Today, (1987), **2.** 339.
96. W. Jones, Catal. Today, (1987), **2,** 357.
97. K. Urabe, J. Sakurai and Y. Izumi, J. Chem. Soc., Chem. Commun., (1986), 1074.
98. J. H. Lunsford, S-L. Fu and D. L. Meyers, J. Catal., (1988), **111,** 231.
99. M. Amara, M. Bettafar, L. Longembre and D. Olivier, Appl. Catal, (1987), **35,** 153.
100. M. Onaka, K. Sugita and Y. Izumi, J. Org. Chem., (1989) **54,** 1117.
101. J. H. Clark, C. V. A. Duke, S. J. Brown and J. M. Miller, J. Chem. Soc., Chem. Commun., (1986), 877.
102. A. Al Jazza, J. H. Clark and M. Robertson, Chem. Lett., (1982), 405.
103. J. H. Clark, C. V. A. Duke, S. J. Brown and J. M. Miller, J. Chem. Soc., Chem. Commun., (1986), 877.
104. S. L. Regen and C. Kotell, J. Am. Chem. Soc. (1977) **99,** 3837.
105. J. T. Kelly, W. Schoen and C. N. Sechrist, U.S. Patent, 3248 343.
106. G. J. Antos, J. C. Hayes and E. L. Pollitzer, U.S. Patent, 3960 711.
107. G. A. Olah and G. Messina, U.S. Patent, 3993 587.
108. T. Ando, T. Kawate, J. Yamawaki and T. Hanafusa, Chem. Lett., (1982), 935.
109. W. P. Weber and G. W. Gokel "Phase Transfer Catalysis in Organic Synthesis," Springer-Verlag, Berlin, (1977), p. 15.
110. M. Tanaka and M. Koyanagi, Synthesis, (1981), 973.
111. M. Hojo and R. Masuda, Tetrahedron Lett. (1976), 613.

3 Analytical Techniques for Studying Supported Reagents

3.1. Introduction

The study of supported reagents by physicochemical techniques has lagged a long way behind their applications in organic synthesis. This is in spite of the enormous amount of research that has gone into more traditional areas of surface chemistry and the large and diverse range of techniques that can be used to study solids and, more especially, solid surfaces. A significant number of articles on the characterization of supported reagents, however, have appeared in recent years and these have generally demonstrated the value of characterization studies in optimizing the reaction performance of supported reagents, as well as understanding their behavior. This chapter is written with the synthetic chemist in mind and is largely biased toward the more common spectroscopic and nonspectroscopic techniques and to their use in improving the reaction properties of supported reagents, as well as helping to determine the best supported reagent for a particular type of application. In this sense it does not duplicate the many texts and review articles written on surface characterization techniques, although some mention is made of more specialized and less available techniques that may yield important information on supported reagents. Examples of supported reagents and supports have been selected to illustrate the value of the various techniques and the ways that they have been applied. The information about a supported reagent that a synthetic chemist is likely to require is:

1. The efficiency of reagent dispersion
2. The nature of the adsorbed species
3. The nature of the support–reagent interaction
4. The effects of changing the support
5. The distribution and strength of active sites
6. The stability of the supported reagent
7. The surface area and porosity of the supported reagent

Table 3.1 summarizes the techniques described in this chapter, as well as the information most commonly or easily obtained by their application and some comment about their general advantages and disadvantages. The reader may use this table as a quick checklist to decide on the most appropriate technique to obtain desired information on a particular supported reagent. The sections thereafter provide more information with selected examples on the applications of these techniques in this field. Readers requiring more detailed information on the methods should refer to specialist texts on techniques or to texts or review articles on surface characterization such as those that have appeared at regular intervals in *Analytical Chemistry*.

3.2. Infrared Spectroscopy

Infrared (IR) spectroscopy is probably the most useful and widely used technique to study supported reagents. IR spectrometers are available in many laboratories and spectra can often be obtained quickly, although more detailed investigations may require the use of more sophisticated equipment. The types of information about supported reagents that IR spectroscopy can provide include: (1) identification of the surface species; (2) dispersion of the reagent over the support surface; and, (3) surface activity studies with the use of probe molecules.

IR can also be used to distinguish different types of surface hydroxyl groups. The major disadvantage of IR spectroscopy in this context is that large regions of the normal IR spectrum are usually obscured by strong bands because of the structural vibrations of the support itself. Raman spectroscopy can often be used as a complementary technique to help overcome this difficulty.

The use of IR spectroscopy to study solid materials and the surfaces of these materials is well documented. The most common ways of studying an insoluble solid are: (1) as a mull; (2) as a disc; and (3) directly as a powder. These methods differ in terms of degree of difficulty (in obtaining useful spectra), ease of sample preparation, and reliability of the information obtained. Table 3.2 gives a summary of the main advantages and disadvantages of these different methods.

Table 3.1

Technique	Information Obtained	Comments
Infrared spectroscopy (IR)	Identification of surface species. Dispersion of reagent over support. Identification of surface hydroxyls. Surface activity via probe molecules.	Versatile and widely available technique. Special equipment (e.g., a diffuse reflectance cell) may be necessary. Support vibrations obscure wide regions of normal IR spectrum. Probably the most useful analytical technique for studying supported reagents.
Ultraviolet–visible spectroscopy (UV–Vis)	Observation of adsorbed reagents or probe molecules. Measurement of adsorption via determination of residual reagent in solution. Measurement of surface solvatochromic parameters.	Rarely used for studying supported reagents. May be useful for transition metal species adsorbed on white supports. Good quality reflectance spectra may be difficult to obtain.
X-ray photoelectron Spectroscopy (XPS) and Auger Electron Spectroscopy (AES)	Analytical information on surface composition. Relative reagent–support binding energies.	Established technique for studying solid surfaces. Not commonly available.
Nuclear magnetic resonance (NMR)	Identification and differentiation of active sites. Interactions between supported reagents and adsorbed molecules. Local environment of constituent atoms. Diffusion of molecules. Mapping of support structures.	Direct studies possible via high resolution solids NMR and wide line NMR.
Electron spin resonance (ESR)	Distinguishing mobile and immobile paramagnetic species.	Rarely used for the study of supported reagents.
Thermal analysis techniques (TG, DSC, DTA, TMA, DIL)	Water content and dehydration. Hygroscopicity. Optimum drying conditions/thermal activation.	Simple and effective technique for measuring hydration, hygroscopicity and thermal stability.

Table 3.1

Technique	Information Obtained	Comments
	Decomposition and phase change temperature.	
X-ray techniques (XRD and EXAFS)	Powder X-ray diffraction is a useful technique for studying; the dispersion of the reagent, the identity of surface species, the basal spacing in clay based reagents. EXAFS can be used to determine the oxidation state of the supported reagent.	Basic dispersion measurements require simple equipment. Basal spacing measurements require more elaborate equipment.
Scanning electron microscopy (SEM)	Reagent dispersion. Size and shape of the particles of the reagent and the support. Element mapping (with X-ray diffraction).	No information on composition. Limited application. Fairly simple technique—useful for quick evaluation of dispersion/presence of microcrystals.
Mass spectrometry (MS, FABMS, etc.)	FABMS and similar soft ionization techniques can be used to provide information on surface composition and reagent loading and dispersion.	Relatively new technique but with much potential. Can be very sensitive and useful for observing complex surface species.
Nonspectroscopic methods	Measurements of acid and base strengths and distribution of these and other sites. Surface area and other structural properties and measurements.—structural damage due to reagents, pore blocking.	Acid and base site strengths and concentration easily measured by titration but limited by range of common indicators. Surface area (etc.) measurements generally use small molecules and may not measure availability to larger reagent and/or substrate molecules.

Table 3.2 Comparison of Different Methods of Studying Supported Reagents by IR Spectroscopy

Method	Advantages	Disadvantages
Mull (nujol, HCBD, etc.)	Simple procedure requiring minimal equipment	May be difficult to achieve good dispersion Mull bands obscure regions of spectrum. Not suitable for detailed studies on the surface groups
KBr–disc (or similar)	Sample procedure requiring commonly available equipment. No bands in the normal IR spectrum due to the diluent	The method can lead to damage of the support–reagent interaction. Generally unsuitable for detailed surface characterization work
Self-supporting disc	No other species present in the analysis. Suitable for most types of study with appropriate cells	May be very difficult to prepare and preparation can damage the material
Diffuse reflectance ("DRIFT")	Most direct method requiring only that the material is powdered. Suitable for most types of study especially with an environmental DRIFT facility (variable temperature and atmosphere)	Low energy throughput generally demands the use of FTIR. Spectral distortion can occur.

Mulling the solid with a suitable matrix liquid, such as nujol or fluolube, is the easiest of the operations and requires the minimum of equipment. This method has been used for the IR analysis of a variety of supported reagents including those based on clays,[1] zeolites, aluminas, and silica gels.[2] Good dispersion of the material in the mull, however, can be difficult to achieve and the mulling agent may obscure spectral regions of interest. The method is only suitable for direct analysis of the bulk material and usually cannot be used, for example, to probe the surface activity or to obtain detailed information on surface hydroxyls.

The majority of articles that describe the use of IR spectroscopy for the study of supported reagents use the disc technique. In the simplest case, the supported reagent is diluted with KBr (or another IR transparent and unreactive diluent) and pressed into a disc[3] following exactly the same procedure as is commonly used for the IR analysis of organic solids. Diluents, such as KBr, however, can damage the support– reagent interaction and are likely to interfere with any in situ studies such as the adsorption of probe molecules. For these reasons, many researchers prefer to use pressed discs of the pure supported reagent ("self-supporting" discs). Stable discs of the supported reagent can be difficult to prepare but a pressure of ca. 20 atmospheres and a wafer concentration of 10–20 mg/cm^2 may often prove adequate. Once prepared, discs of this type offer no interference from foreign species, and can be treated in situ (e.g., thermally to remove water) and used for subsequent in situ adsorption studies of probe or reactant molecules.

Both the mull and disc methods are transmittance techniques. IR opaque or highly scattering materials may not be suitable for analysis by transmission spectroscopy. The disc method is also not suitable for thermally sensitive materials (e.g., where the dispersed reagent is subject to thermal decomposition) due to the high temperature generated in pressing the wafer. In cases such as these the diffuse reflectance method can be a very useful alternative. This involves irradiating a powdered sample, then measuring the spectrum of a nonabsorbing standard such as KCl or KBr. Indeed, the only restriction to the use of this method is that the sample must be powdered, although in practice, the low energy throughput associated with the method effectively means that the rapid scanning and high energy provided by a Fourier transform instrument is required. Several types of commercial diffuse reflectance accessions are available including those incorporating environmental chambers that enables the material under investigation to be studied at different temperatures and under different atmospheres. These options can be very useful to study thermal stability and to probe molecule studies involving the spectroscopic titration of surface sites (discussed later).

The diffuse reflectance method is generally a very convenient technique to study supported reagents. One of its major advantages is that it can often be applied to the pure, untreated material. Spectral distortion can, however, occur when observing pure materials. This may even take the form of extra bands in the spectrum, as shown in Figure 3.1, for $Na_2[Cu(CN)_3]$[4]. An extra band also appears in the spectrum of $Na_2[Cu(CN)_3]$–alumina. Fortunately, these bands can be removed by simply diluting the samples with a nonabsorbing matrix such as KCl or KBr.[4] Indeed, the loss of spectral distortion of dilution suggests that specular reflection (in addition to the desired diffuse reflection) is causing the distortion. Specular reflection that is due to reflection from the front surface of the sample, normally appears as "Reststahlen" bands that show complete inversion of normal absorbance bands in regions of intense IR activity and that are also slightly shifted to higher wave numbers.[5]
Diffuse reflectance is generally a qualitative IR technique; this is another important associated fact. It can be used as a quantitative technique if the sample and the dilution matrix (if any) meet certain criteria that came from the assumptions made in the Kubelka–Munk equation.[4] This is the

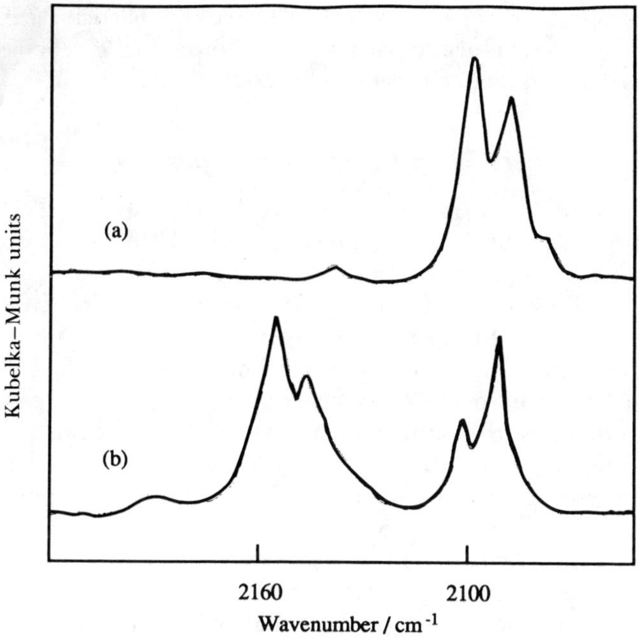

Figure 3.1 Extreme distortion effects in the "DRIFT" spectrum of $Na_2[Cu(CN)_3]$ due to (in part) specular reflectance[4] (a) $Na_2[CuCN)_3]$ diluted to 1% by weight with NaCl; (b) undiluted $Na_2[Cu(CN)_3]$.

diffuse reflectance analogue to Beer's law for transmission measurements and is valid for samples in which an increase in depth does not appreciably change the spectrum. Unfortunately, this may not be the case for well-dispersed supported reagents where the thickness of the dispersed reagent is at the molecular level. The Kubelka–Munk function is defined as:

$$\frac{K}{S} = \frac{(1 - R^{\infty})^2}{2R^{\infty}}$$

where R^{∞} is the diffuse reflectance of the sample; K is the absorption coefficient; and S is the scattering coefficient. For a sample to show the ideal linear relationship between F and concentration, the following criteria must be met:

1. There must be negligible specular reflectance (discussed earlier)
2. The sample must be dilute (e.g.,, by using an inert matrix such as KBr)
3. S should be constant both as the concentration is varied and across the whole spectrum

Both scattering and specular reflectance are affected by particle size, therefore, a uniform and reproducible particle size is important if reliable quantitative information is to be derived from spectra.

3.2.1. Direct Characterization of Surface Species

In spite of its obvious potential as an analytical technique for characterizing absorbed species, attempts to use IR for this purpose can be frustrated by significant regions of the IR spectrum being totally masked by the absorptions of the support itself. This effectively denies the use of a large part of the so-called fingerprint region of the IR spectrum and obviously makes the direct observation of many molecules very difficult. Nonetheless, there are many reagents of interest that possess IR active groups that absorb in other quiet parts of the spectrum and for these IR can be the technique par excellence for molecular characterization and to study the interaction between the support and the reagent.

In the simplest case, IR can be used to confirm quickly the presence of the expected species. Thus, IR has been used to confirm the presence of cyanide ions in a NaCN–alumina reagent by using KBr pellets.[6] Similarly, the retention of alkyl ligands in silica-supported diallylrhodium was readily proven by IR using a self-supporting disc.[7]

Adsorption of a reagent onto a support can cause significant changes in the IR spectrum of the reagent. Thus, the IR spectrum of a freshly prepared silica–$Cu(OCOCH_3)_2$-H_2O reagent shows a shift to lower energy in the position of the antisymmetic CO_2-stretching mode of 30 cm^{-1} compared with that of the free reagent.[8] Thermal treatment of the supported reagent ultimately resulted in the loss of these bands suggesting the loss of acetate (as acetic acid). Exposure of the resulting material to acetic acid vapor restored the original spectrum.

Changes in the IR spectrum of the reagent on adsorption can also help to explain changes in reaction behavior. Silica-supported $KMnO_4$ is known to be a useful oxidant and can show different a chemistry to that of unsupported $KMnO_4$. The supported reagent can actually be fine-tuned to give maximum selectivity in some oxidation reactions by the use of IR spectroscopy. Careful optimization of the reagent loading and the drying temperature were shown to be important, therefore, under the correct conditions, the IR spectrum of $KMnO_4$ showed appreciable distortion of the reagent consistent with a change from T_d to C_{2v} symmetry (Fig. 3.2.).[9]

Simple mulling of supported reagents can sometimes be sufficient to enable IR monitoring of changes in the surface-bound species. This is especially true when only a qualitative indication of a fairly major change is involved. Thus, the conversion of $Rh(OAc)^{x+}{}_{4-x}$ in a hectorite to $Rh(PPh_3)_n{}^+$ can be easily witnessed by the replacement of IR bands from the acetate by those from the phosphine.[1]

In cases where spectral changes are expected over a wide energy spec-

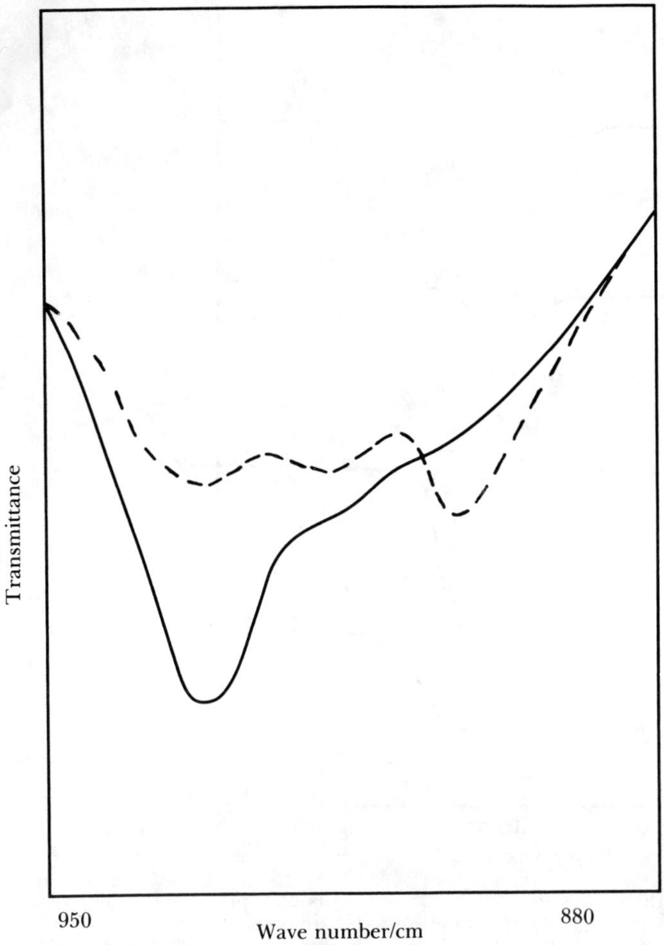

Figure 3.2 IR spectra of silica-supported $KMnO_4$ before drying (– – –) and after drying at ca. 100°C to remove physisorbed water.[9]

trum, the complementary use of IR and Raman spectroscopy can be useful. Silica pellets of the material show an IR spectrum consistent with the presence of alkyl groups.

Treatment of this supported reagent with H_2 caused a reduction in intensity of the bands attributed to the allyl groups accompanied by a growth in bands, which have been ascribed to the Rh–H products shown previously.[10] The intense IR absorption below 1400 cm^{-1} due to the silica support prohibited the use of IR for the study of other characteristic bands, but Raman spectroscopy revealed bands due to Rh–O vibrations. The Raman spectra were obtained by using a Kr laser (6764 Å) on powdered samples contained in a glass capillary tube.

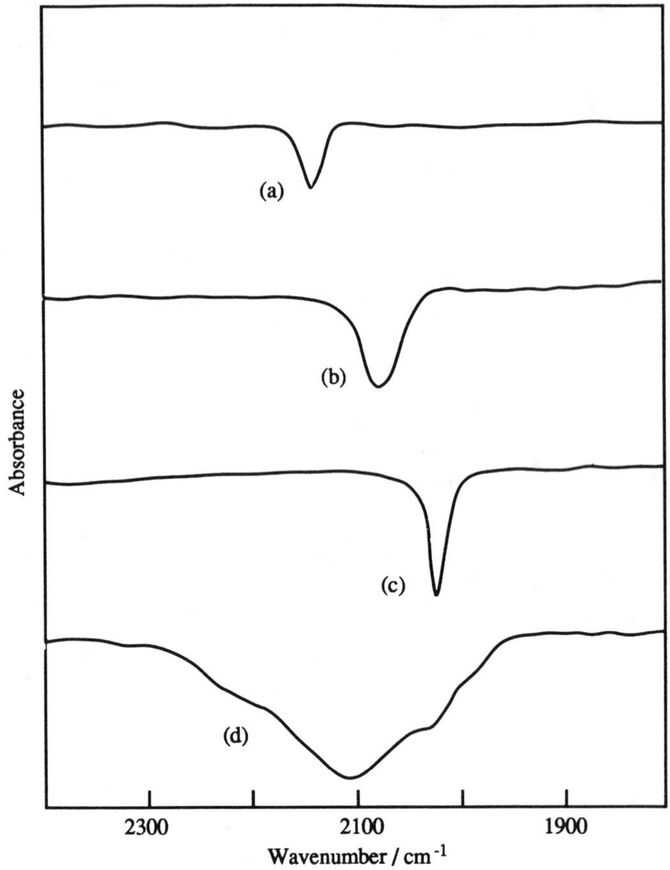

Figure 3.3 Shirts in the $\upsilon(C{\equiv}N)$ band of CuCN and KCN on supporting onto alumina (shifts in the same direction but of different magnitudes are observed with different support materials). (a) CuCN; (b) CuCN–alumina (0.9 mmol/g)$_J$ (c) KCN; and (d) KCN–alumina (0.9 mmol/g).[13]

Supported cyanides have been the subject of considerable attention and the development of a number of useful synthetic reagents. This is a consequence of parallel IR investigations directed toward optimizing the loading (discussed later) and the identification of the surface species. The $\upsilon(C{\equiv}N)$ band of CuCN undergoes a downfield shift on binding to various support materials (Fig. 3.3)[11] in contrast to an upfield shift on supporting KCN on the same materials (Fig. 3.3).[12] The observed position of $\upsilon(C{\equiv}N)$ in supported CuCN is consistent with the formation of $Cu(CN)_2^-$ and this suggests that the unique reactivity of such supported reagents in nonpolar solvents[11] may be due to the presence of Cu^+, presumably stabilized by support hydroxyls. Furthermore, IR studies on these materials gave information on their thermal stability and the apparent formation of $Cu(CN)_3^{2-}$ species at higher reagent loadings. The optimized methods of preparation, support type, drying temperatures, and reagent loading as predicted by the IR studies were used to develop the most reactive and efficient supported CuCN reagents.[11,13]

The $v(C{\equiv}N)$ vibration is an excellent example of a band that occurs in a quiet region of the IR spectrum and is sensitive to adsorption, loading, and the nature of the adsorbed species. Its value has been further exploited in the study of supported copper(I) thiocyanate[14] where IR can be used to (1) distinguish between thiocyanate and isothiocyanate; (2) determine the best method of preparation; and (3) help determine the identity of the adsorbed species.

Research of this type clearly shows that IR can be a quick and reliable technique to screen supported reagents in terms of their probable reactivity and selectivity. IR can also be useful as one weapon in an armory of techniques applied to supported reagents that are less amenable to a single type of analysis. One good example of this type of reagent is KF–alumina, which has been the subject of many forms of analysis and where composition and source of reactivity remains a matter of controversy. Variable temperature IR studies on KBr discs of KF–alumina clearly show the loss of physisorbed water but also reveal the formation of AlF species indicative of the reaction of support and reagent.[15] Interestingly, the formation of carbonate is also apparent[16] and this has been found to result from the reaction of thermally activated KF–alumina with air.[17] This may indicate the formation of air-sensitive surface O^{2-} sites, which may help account for some reports of highly basic forms of this material. IR has also been used to reveal the formation of Si–F-containing species in silica-supported fluorides.[18]

IR studies, such as those on supported fluorides, have helped to show that IR can be a useful technique even when the major spectral changes occur in the busy low energy region of the spectrum. Subtle changes in this region of the spectrum have been the subject of many articles, although useful information often can only be obtained by pain-staking research and the use of nonroutine equipment. Changes in the region for example, have been found to be useful in the characterization of pillared and unpillared clays containing a variety of interlamellar ions.[10] Such work is typically carried out using self-supporting discs and specially designed, evaluable cells. In this way, it is possible to identify structural groups such as Al-OH-Al and Al-OH-Mg. It can also be used readily to observe the loss of ammonia resulting from the NH_4^+ activation of clays. Cation-dependent spectral changes are often observed.[19]

High temperature heating (to 700°C+) brings about major changes in the spectra of all clays. These are due to the loss of hydroxyl groups and to changes in the coordination of the Al and Mg species in the central layer of the clay sheet. Variable temperature IR can also reveal important information about the clay pillars such as the diffusion of protons from the pillaring polymer (e.g., Al_{13} species) during calcination.[19]

3.2.2. Studies on Reagent Dispersion

IR-active reagents that show a band or bands in a quiet reagent of the spectrum can be excellent candidates for dispersion studies as adsorption onto a solid support is often accompanied by some change in the spectrum. This has been effectively exploited in many cases including supported permanganate,[9] KCN,[20] CuCN,[13] CuSCN,[14] cyanide complexes,[4] KSCN,[15] and NaN_3.[2] Most of these reagents have diagnostically useful IR bands.

Azides (in common with other species containing either cumulative double bonds or isolated triple bonds, e.g., CN^-, SCN^-, CO_2, allenes, acetylenes) have a relatively strong, sharp infrared band in the region of

Figure 3.4 IR spectra of NaN_3 supported on alumina showing the effects of loading and water contect on the $v(N_3^-)$ band.[2] (a) NaN_3; (b) NaN_3–alumina (ca. 0.7 mmol/g) containing 21 wt % H_2O; and (c) NaN_3–alumina (ca. 1.4 mmol/g) containing 7.0 wt % H_2O.

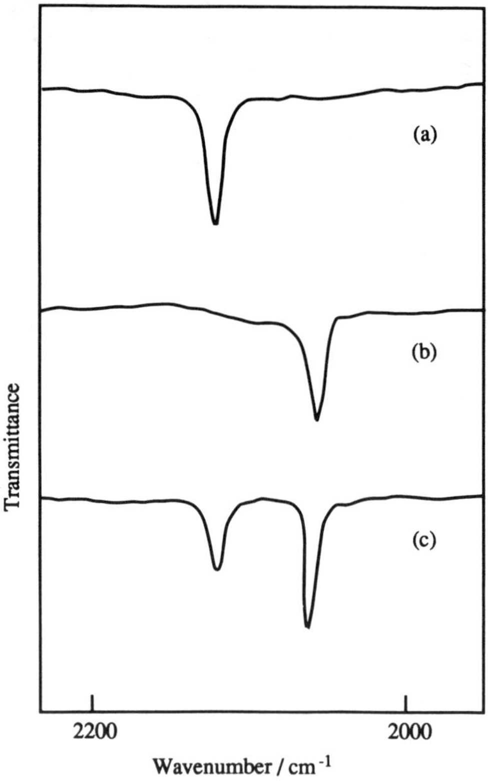

2100–2200 cm^{-1}. This is in a quiet part of the spectrum, far away from the broad intense bands of most supports, and allows for good spectra to be recorded using conventional techniques (the authors used a Nujol mull in the present example).

Sodium azide itself has a band at 2130 cm^{-1} (Fig. 3.4). The optimally loaded Al_2O_3-supported reagent (0.76 mmol/g) has one sharp band at 2060 cm^{-1} with no peak at 2130 cm^{-1}. This implies that all the azide interacts with the alumina surface, with no crystalline sodium azide present. The relative intensities of the free and supported azine ion stretching bands are dependent on loading and water content. This reagent gives a 74% yield of product. Increasing the loading to 1.6 mmol/g reduces the yield to 47%. The infrared spectrum of this material, however, shows a large band at 2130 cm^{-1}, probably corresponding to microcrystals of sodium azide resulting from full surface coverage by ~ 50% of the azide present, forcing the other half of the azide to form less well-dispersed clusters of sodium azide. Analysis of the spent reagent from a reaction of the overloaded material shows that the band at 2060 cm^{-1} has almost disappeared, whereas the 2130 cm^{-1} band is essentially unchanged. This proves that the active species in this supported reagent is at 2060 cm^{-1}, the only species present in the optimally supported reagent. Further discussion on the use of infrared spectroscopy to help improve the performance of supported reagents is given in Chapter 2.

Reaction of the supported azides clearly showed the supported rather than the free ions to be the most reactive.[2] Very similar and equally informative studies have been carried out on supported KCN,[20] which also show a support-dependent $v(CN^-)$ band (Fig. 3.5) that can be related to the observed support-dependant reactivity. (KCN–alumina is highly reactive, whereas KCN–silica, for example, is relatively unreactive presumably due to a reaction between the silica and the KCN causing an especially large shift in the v(CN) band.)

3.2.3. Studies via Probe Molecules

One way around the problem of studying a supported reagent that is not itself amenable to IR analysis is to use suitable probe molecules that can "spectroscopically titrate" the active sites. These probes can be selected to match the supported reagent under investigation, therefore, in the case of acidic materials, amines are excellent probe molecules. All that is required of the probe is that its IR spectrum is sensitive to interactions with the types of site (acidic, basic, oxidative, reductive, coordinative, etc.) assumed to be present on the material.

Both Lewis and Brönsted acid sites were identified on the surface of various inorganic oxide supports treated with Al_2Cl_6[21] by using pyridine as the probe molecule and the magnitude of the shift, which ranged from 8–18 cm^{-1}, was found to be support dependent. Silica and high silica

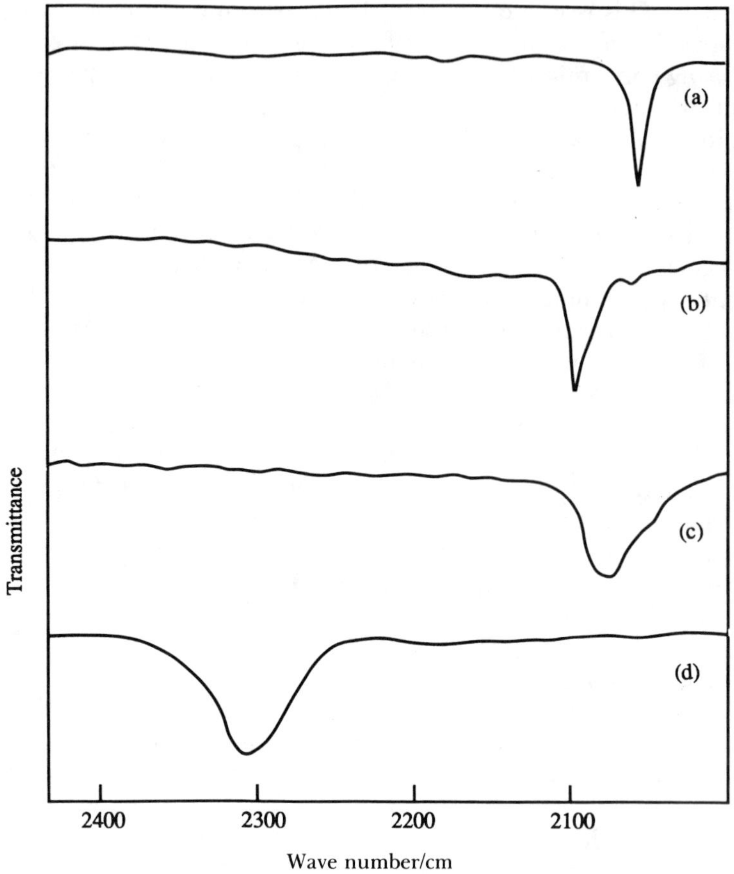

Figure 3.5 IR spectra of KCN and supported KCN showing the effects of loading on the $v(CN^-)$ band.[13] (a) KCN; (b) KCN–charcoal (6 mmol/g); (c) KCN–alumina (0.9 mmol/g); and (d) KCN–silica (3 mmol/g). Loadings are based on theoretical monolayer formation.

zeolites gave the largest IR band shift. The measured activity correlated well with the reactivity of these materials in typically acid-catalyzed reactions. Similarly, ammonia adsorption has been used to study the surface acidities of γ-alumina exposed to a number of reagents including $POCl_3$, PCl_5, and H_2PO_3F.[22] For volatile reagents, such as $POCl_3$, self-supporting alumina wafers were first burned in a stream of air at 823 K followed by admission of the reagent vapor in the IR cell at 573 K. The resulting materials were then exposed in situ to gaseous ammonia at 298 K and spectra were recorded at this stage and after ammonia desorption at 373, 473, and 573 K. For less volatile reagents, a vessel containing the reagent was placed in the heated zone of the cell and volatilized onto the alumina wafers. The adsorption of ammonia was found to result in substantial changes in the IR spectrum with several new bands appearing, some of which survived subsequent heating. The location of these bands enabled the assignment of Lewis and/or Brönsted acid sites (bands at ca. 3200, 3100, and 1400 cm^{-1} are due to NH_4^+ and can be attributed to reaction of NH_3 with Brönsted sites).

The strength of adsorption of probe molecules can provide important information relating to surface-site activities and to the identification of reaction sites. It may actually be possible to use a probe molecule to specifically react with certain surface sites. Thus, in the case of silver-impregnated silica gels, for example, IR monitoring of adsorbed iodomethane was to probe the amount of ion-exchanged Ag^+ ions (which caused irreversible adsorption of CH_3I, presumably due to formation of AgI) as opposed to nonion-exchanged Ag^+ ion (which caused reversible adsorption of CH_3I).[23]

Probe molecules that have IR adsorption in quiet regions of the spectrum can be used to easily and quickly reveal important information about the surface species present in supported reagents. Perhaps the two best examples of such molecules are CO and organic nitriles, which both show strong absorptions in the quieter, higher energy region (>1600 cm^{-1}) of the IR spectrum and are sensitive to adsorption on metal ion sites.

Thus, IR has been used to study the adsorption of CO on copper–silica catalysts containing Cu^{2+}, Cu^+, and Cu^o.[24] In the case of unreduced samples containing Cu^{2+} (prepared by ion exchange of an aerosil-type silica) two clearly resolved CO adsorptions at ca. 2170 and 2199 cm^{-1} were observed. The former band has been attributed to small CuO particles, whereas the latter is thought to be due to isolated Cu^{2+} ions. After catalyst reduction, adsorbed CO shows bands at 2090–2113 cm^{-1} and 2175 cm^{-1}, and these have been attributed to the presence of small Cu metal clusters and isolated Cu^+, respectively.

Similar studies on the adsorption of CO onto alumina-and silica-supported iron catalysts have also helped to reveal the presence of different metal sites.[25] In the case of $Fe–Al_2O_3$ (prepared by supported ferric nitrate), for example, reduction of the catalyst under increasingly severe conditions caused changes in the IR spectrum of adsorbed CO (Fig. 3.6). Interestingly, the band at 2155 cm^{-1} is unaffected even under quite severe reduction conditions (873 K for 2 hr) and this has been attributed to the presence of Fe^{2+} cations, which are strongly interacting with the support and, therefore, difficult to reduce. The increase in intensity of the other bands in the CO spectrum at 2050 and 2010 cm^{-1} on increasingly forcing reduction conditions strongly suggest that they are indicative of metallic iron atoms.

Various clay-based supported reagents have been studied by using organic probe molecules having strong, characteristic IR absorptions that are sensitive to metal ion–organic probe interactions. The $v(C{\equiv}N)$ band of benzonitrile, for example, is an apparently useful probe of the Lewis acid activity of various montmorillonite–metal halide-supported re-

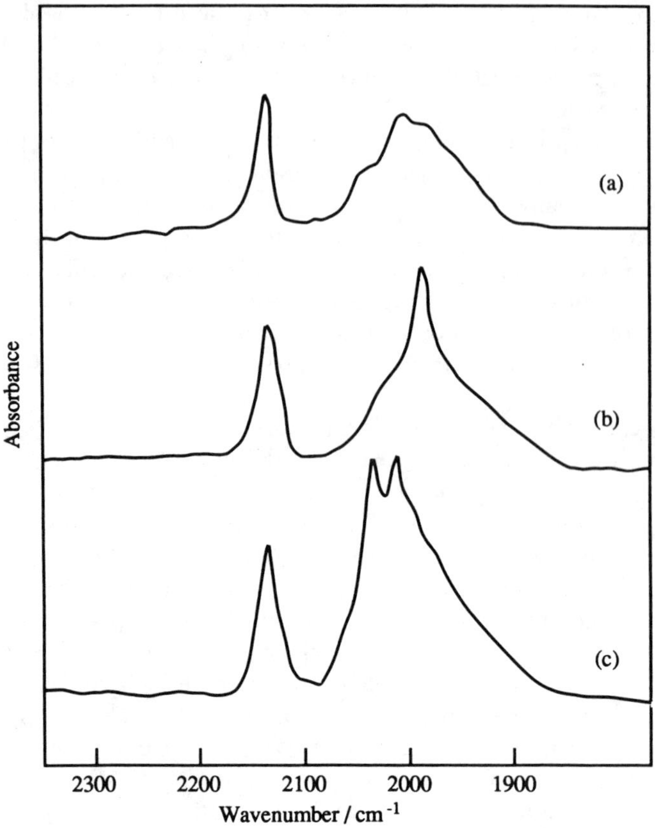

Figure 3.6 IR spectra of absorbed CO on γ-alumina-supported iron (prepared from alumina–$Fe_2(NO_3)_3$ that has been reduced at 773 K for 0.5 hr (a); 873 K for 0.5 hr) (b); and 873 K for 2 hr (c).[25]

agents.[26] The observed shift to higher frequency in this band relative to free benzonitrile (Table 3.3) correlates quite well with the observed reactivity of the catalyst in Friedel–Crafts reactions.[27]

3.3. Electron Spectroscopy

3.3.1. Ultraviolet–Visible Spectroscopy

Ultraviolet–visible spectroscopy (UV–VIS) has received little attention as a technique to study supported reagents. Apart from the trivial use of the technique to study excess reagent in solution (and hence determine loading) or desorbed reagent (and hence determine the identity and quantity of reagent released by, for example, washing of the supported reagent), reflectance equipment is required to obtain a spectrum

Table 3.3 Observed Shifts in the $v(C{\equiv}N)$ Band of Benzonitrile on Adsorption onto Montmorillonite (K10)–Metal Halide-Supported Reagents[26]

Supported Reagent	$\Delta v(C{\equiv}N)$PhCN
K10	0
K10–$FeCl_3$	23
K10–$CoCl_2$	29
K10–$ZnCl_2$	32
K10–$NiCl_2$	34
K10–$CuCl_2$	43

of the adsorbed reagent. Commercial equipment may also cover the near infrared region of the spectrum. The most likely applications of reflectance UV–VIS in this area are:

1. To observe the electronic spectrum of adsorbed reagents
2. To observe the electronic spectrum of adsorbed reactants or products
3. To use adsorbed spectroscopically active molecules as probes of surface activity

In principle the techniques is similar to reflectance IR, although in practice it is less widely used and may suffer from drawbacks including:

1. Difficulties in obtaining good quality reflectance spectra
2. Broad, often overlapping, bands in the UV–VIS (electronic transition) region

Changes in the oxidation states of adsorbed transition metal complexes may be conveniently observed by reflectance UV–VIS spectroscopy, if the support itself (or any other species present) does not cause adsorption in the same region. Thus, the aerial oxidation of Co^{2+}–hectorite (a white clay due to very low structural iron content) treated with bipyridyl gives adsorbed $[Co(bipy)_3]^{3+}$ as shown by the adsorption spectrum.[28] The spectrum was obtained by sandwiching the sample between quartz plates with magnesium oxide as the reference.

The nature of the active sites of several supported $KMnO_4$ reagents has been determined by UV–VIS spectroscopic titration of basic and acidic sites with indicators such as pyridine and *m*-dinitrobenzene.[29]

A more unusual but very interesting application of reflectance UV–VIS in the analysis of solid supports is in the measurement of the solvatochromic parameter π^*. This parameter is normally associated with solvents and is considered to be a measure of polarity and/or polarizability of the liquid. By measuring the position of the adsorption maximum in the reflectance spectrum of adsorbed *p*-nitro-anisole, π^* has been measured for a series of silica surfaces. Remarkably, whereas ordinary solvents have π^* values ranging from 0.00 (for cyclohexane) to 1.09 (for water), the values measured for eight different silicas were in the range of 2–2.8, suggesting remarkably high surface "polarities."[30]

Near infrared and visible spectroscopy has been used to observe the overtone and combination vibrational spectra of adsorbed water and ammonia and the electronic spectra of silica-supported Ni(II) reagents prepared by different methods.[31] The study provided useful information about the local environment of the transition metal ion including the geometry of adsorbed Ni(II) complexes and the ease of solvation of the metal. The latter depended on the method of preparation of the supported reagent so that Ni(II) adsorbed on calcined silica or by ion exchange was less easily solvated by H_2O or NH_3 compared with ordinary impregnated samples. This suggests that a stronger interaction occurs between nickel and $Si{-}O^-$ (from calcination or by ion exchange) than nickel and Si–OH sites as might be expected.

3.3.2. X-ray Photoelectron and Auger Electron Spectroscopy

X-ray photoelectron (XPS) and Auger electron spectroscopy (AES) are well-established techniques for the analysis of solid surfaces[32,33] primarily giving information on surface composition by direct excitation of specific atoms. Unfortunately, the unavailability of the necessary equipment in many laboratories has severely restricted the application of these powerful techniques to the analysis of supported reagents.

The results from an XPS study of silica-supported copper(II) materials along with those from SEXAFS (see Section 3.6.2) have been used to determine the nature of the active sites on the surface.[34] Generally the binding energies for metal oxides in supported reagents tend to be higher than those of metal oxides themselves and this is usually ascribed to the coordination effects of supports or to the formation of new compounds.[35,36] Based on this observation, it appeared that copper(II) acetate interacted more weakly with the silica support than a tetraamine copper(II) complex. Surface compositions of different supported reagents were easily obtained by measuring the intensities of the peaks due to copper and silicon.

The reaction of silica-chemisorbed diallyrhodium with H_2, to give highly dispersed rhodium metal complexes has been investigated by XPS.[7] The results showed a shift of the Rh third levels to lower binding energies consistent with the reduction of the supported rhodium. Reoxidation of Rh(O) to Rh(III) by treatment with HCl was also easily observed by a shift in the XPS spectral peak.

Thus, both strength of interaction with the support and changes in supported reagent oxidation state can be monitored by XPS. A wider use of this essentially simple and direct (if relatively uncommon) technique to the study of supported reagents would seem to be appropriate.

3.4. Magnetic Resonance

3.4.1. Nuclear Magnetic Resonance Spectroscopy

Nuclear magnetic resonance (NMR) spectroscopy is now established as one of the most powerful of analytical techniques and its application to the analysis of supported reagents is increasing in popularity. This is largely due to the advent of commercial high-resolution solids NMR spectrometers that use magic angle high-speed spinning, high-power decoupling, and cross-polarization techniques to reduce the very broad NMR resonances, which are normally associated with solid samples.

NMR spectroscopy can be used to provide information on

1. The identification and differentiation of active sites either directly or by the use of probe molecules
2. The interaction between the supported reagent and adsorbed molecules
3. The local environment of constituent atoms including the selective observation of surface species
4. The diffusion of molecules through porous solids

Supported reagents based on all the common support materials have been studied by NMR spectroscopy, although zeolites have attracted the most attention and have been the subject of some of the most sophisticated NMR techniques. Remarkably, high-resolution solid NMR spectroscopy is being used to map out the structures of zeolites at a level of detail and accuracy that may soon rival X-ray diffraction, but without the need to grow single crystals. Although such applications require state of the art equipment and complex pulse sequences, less sophisticated equipment and experiments available on a more routine basis are increasingly used to obtain information on supported reagents. Most work has focused on nuclei commonly present in supported reagents (^{29}Si, ^{27}Al, and ^{1}H) and the more popular NMR nuclei (^{13}C, ^{31}P, and ^{19}F), although in principle at least, most nuclei can be studied by solids NMR.

3.4.1.1. Direct Characterization of Surface Species. Proton magic angle spinning (MAS) NMR has been used extensively in recent years to characterize the surface hydroxyls on various support materials.[37–39] The suppression of proton–proton dipolar interactions (which contribute toward line broadening) can be readily achieved by line-narrowing multipulse sequences as these dipolar interactions are weak enough to be averaged out with sample rotations of only a few kilohertz (which can be achieved with most commercial MAS NMR instruments).[37]

Perhaps the most important information that can be obtained in this way concerns the different acidities of surface hydroxyl groups, which are, of course, present on most support materials. The combined rotation and multiple-pulse spectroscopy (CRAMPS) methods has been used to study the ^1H NMR of silica–alumina, silica gel, and γ-alumina support materials.[40] Sufficiently good resolution was achieved to enable distinction of several different types of hydroxyl species and by studying samples pretreated at different temperatures. It proved possible to identify these species. Rather curiously the δ_H peak at 7 ppm, which is normally associated with a catalytically active Brönsted site, is significantly reduced in intensity on heating samples of silica–alumina. This may indicate that this peak is, in fact, due to hydrated Brönsted sites (1) and that the most active Brönsted sites are, in fact, due to the corresponding "naked" Brönsted sites (II) that must have a much higher chemical shift value.

I II

In general, it is useful to distinguish between the various types of surface hydroxyls, such as terminal bridging and hydrogen bonded, but this may prove difficult even with CRAMPS-type methods. This is due to the small chemical shift range of ^1H, which means that very narrow signals may be required. Careful examination of the spinning-sidebands in terms of their intensities (which indicate anisotropic dipolar interactions[38,39]) or the effects of partial deuteration (which helps distinguish homonuclear from heternuclear dipole interactions, as discussed previously) can help. Further assistance in the assignment of the different ^1H may be obtained by ^1H–^{29}Si cross-polarization (CP/MAS) and, more elegantly, by an extension of that technique to two-dimensional (2D) spectroscopy.[41] 2D CP/MAS ^1H–^{29}Si NMR has been successfully used to identify the protons that are rigidly bound to silicon atoms in silica and zeolite supports.[41]

MAS ^1H NMR is also a powerful tool for the determination of the Al–H distance in rigid polycrystalline solids, such as zeolites, because the second moment (measured as the root-mean-square width) of the NMR resonance signal is inversely proportional to the sixth power of the internuclear distance between nuclei coupled by the dipolar interaction. In this way, the mean distance between acidic protons is bridging hydroxyl groups and the nearest framework Al atom in the zeolite H-Z5M-5, for example, has been calculated as 2.48 ± 0.04 Å[42]

Applications of NMR to the characterization of surface-adsorbed species are as yet somewhat less developed than the NMR characterization of the surface of the support itself. In many cases it should be possible to apply the NMR techniques described earlier to the study of the effect of adsorbed species on surface hydroxyl groups, although relatively little work in this area has been reported.

The obvious alternative approach to the NMR characterization of adsorbed species is to study nuclei contained within the adsorbate itself. Arguably the simplest and most extensively structured nucleus in the solid state is ^{13}C, although its low natural abundance usually means that ^{13}C-enriched samples are required for the study of surface-bound (and hence dilute) molecules.

Static-sample solid-state NMR of adsorbed species is quite well established and was first applied to the study of methanol and its reaction chemistry on the zeolite Z5M-5 in 1978.[43] The broad-line spectrum gave limited information and did little more than corroborate more detailed results from the characterization of reaction species by gas chromatography. The advent of commercial high-resolution solid-state NMR instruments has again revolutionized this area of research so that MAS ^{13} CNMR has been used recently to identify no fewer than 29 different organic species in the same system.[43] The exciting possibility of following reactions on supports is now becoming a reality.

Multinuclear NMR considerably broadens the scope of the technique. Wide-line (non-MAS) NMR continues to find applications especially when the nucleus under observation has a very wide chemical shift range. Thus, ^{63}Cu NMR has been used to study the dispersion of silica-supported copper catalysts.[42] Dispersions were calculated by measuring the integrated intensities of the well-resolved peaks due to copper(I) (assumed to represent all the surface copper atoms) and copper metal (assumed to represent the bulk copper atoms). The calculated dispersions agreed well with those determined by the traditional nitrous oxide method.

Supported fluorides have been the subject of considerable interest in the literature and many characterization methods have been applied to their study. Perhaps the most ambitious analysis was based on MAS ^{19}F NMR,[44] although ^{19}F NMR has many points in its favor (high NMR receptivity and a wide chemical shift range) and is one of the most popular NMR nuclei in the solution state, little is known about it in the solid state because of the limited availability of suitable NMR probes. Interestingly, solid-state ^{19}F NMR proved to have a wider and more informative chemical shift range for fluorides compared with the solution state spectra. KF–alumina, for example, gave a ^{19}F chemical shift very different from those of KF itself and KF in water, and its position may indicate that the fluoride ion is coordinatively unsaturated, which would, therefore, provide vacant sites for attack by incoming substrate molecules. The large difference in chemical shifts between supported and unsupported metal fluorides[44] may also be related to the lack of true F^- reactivity (as opposed to general base activity) in the former. More recent work highlights the importance of Al–F-containing species formed by the reaction of F^- with the alumina and it now seems more likely that the observed ^{19}F signal is due to these species rather than F^- per se (see Section 4.1). In contrast, the reagent KF–CaF_2, which is known to be a good source of F^-, shows a ^{19}F signal very close to that of KF itself, strongly suggesting that this reagent is simply made up of well-dispersed KF on the CaF_2 surface.

3.4.1.2. Studies of Surfaces by Probe Molecules. The use of small molecules as probes of sites on surfaces is a popular method that lends itself to many spectroscopic as well as nonspectroscopic techniques, including NMR. Again the high dilution of surface-bound species is an important factor and makes the direct study of probes containing carbon and nitrogen relatively difficult and time-consuming. Thus, probe molecules, such as ammonia and pyridine, which are popular within the context of IR studies, are less obvious candidates for NMR work although amines have been used to help characterize catalyst surfaces by ^{13}C and ^{15}N NMR.[45–51]

A more attractive nucleus for such studies is ^{31}P, which enjoys high natural abundance, high NMR sensitivity, and short relaxation times. For example, trialkylphosphines have proven to be useful and convenient probes to study acidic sites on amorphous silica–alumina. [45–51] Brönsted and Lewis sites on the surface can be distinguished and variation of the surface phosphine concentration yields a titration of acidic sites. Although the different types of sites can be distinguished and their relative affinity for the probe determined (Brönsted acid sites selectivity remove the first phosphine molecules), it is intersting to note that different phosphines count different numbers of Brönsted sites. This conveys the importance and general message that it is important to use more than one probe molecule in spectroscopic titration techniques of this type.

3.4.1.3. Other Applications of NMR. High-resolution NMR can be a valuable method to elucidate structural aspects of supports and supported reagents and may ultimately rival or even supercede X-ray crystallography (especially as the latter requires the preparation of a crystalline sample), although the subject is still in its infancy. In the case of clays, for example, it has already been demonstrated that NMR can provide useful information on[52]:

1. Distinguishing tetrahedrally and octahedrally coordinated Al
2. The determination of the Si:Al ratios in the tetrahedral sheets
3. The validity of Si, Al distribution models

Wide-line NMR is a useful technique to study the motion of adsorbed molecules.[53–55] In general, line narrowing can be considered as evidence for motion of such species and more specifically a complete narrowing of the line-width to the field inhomogeneity limit (instrument-dependent) results from translational rather than rotational motion. This correlation has been used to study the degree of interaction between the support and the adsorbed species so that in the case of SbF_5 intercalated in graphite, for example, extreme line narrowing of the ^{19}F NMR spectra has been interpreted as indicating effectively "free" SbF_5 molecules.[53] A similar approach has been used to study the molecular self-diffusion of fluorine-containing molecules adsorbed on zeolites[54] and this has revealed a stronger interaction of CHF_2Cl compared with CF_4. The relative self-diffusion coefficient of different molecules in different zeolites, as measured by NMR, correlates well with the diameters of the molecules and of the windows of the zeolites as might be expected.

3.4.2. Electron Spin Resonance

Electron spin resonance spectroscopy (ESR or EPR, electron paramagnetic resonance) has received relatively little attention, compared with NMR, as a technique to study supported reagents, although it can obviously be useful for those materials that contain paramagnetic species.

ESR allows the distinction between mobile and immobile paramagnetic species, for example, which can be very useful in the study of charged supported reagents such as metal ion doped clays and zeolites.

ESR has been used to show the differences in the rotational mobility of Cu^{2+} inside montmorillonite clays that contain different sized complexing solvents.[56] As small a change as from methanol to ethanol can result in a change from a freely rotating Cu^{2+} center located in the clay interlayer, to a rationally restricted Cu^{2+} center. The latter is easily recognized by an anisotropic ESR spectrum. Copper-doped alumina-pillared montmonillonite clays have also been extensively studied by ESR,[57,58] which helps to reveal the nature and location of the Cu^{2+} ions. The analysis has often revealed the adsorption of the ions on to the pillars and this interaction may be irreversible.

3.5. Thermal Analysis

Thermal analysis covers a series of techniques that determines the physical and chemical properties of a substance as a function of temperature and/or time. The major techniques available are:

TG — Thermogravimetry (measurement of changes in weight) including temperature-programed desorption.

DSC/DTA — Differential scanning calorimetry/differential thermal analysis (determination of energy changes during physical transitions and chemical reactions)

TMA	Thermomechanical analysis (measurement changes in material length under a defined load)
DIL	Dilatometry (determination of dimensional changes of materials when subject to a controlled temperature program)

Simultaneous thermal analysis (STA) is usually a combination of TG with DSC/DTA. Evolved gas analysis (EGA) involves the coupling of thermoanalytical instruments (most commonly TG) with a mass spectrometer, gas chromatograph, or infrared spectrometer to facilitate the analysis of gaseous reaction products. Although TMA and DIL techniques can provide important information on, for example, granulated catalysts that may be required for fixed bed column reactors, TG and DSC/DTA (along with EGA) are the most useful routine thermal analysis techniques for the study of supported reagents.

These techniques can easily and quickly provide useful information on:

Water content, dehydration and hygroscopicity

Optimum drying conditions/thermal activation

Decomposition and any phase change temperatures

Very small quantities (typically $\not>$ 10 mg) are required and the analyses are carried out on the neat supported reagent. More elaborate studies involving controlled atmospheres are possible.

3.5.1. Thermogravimetry (TG)

The temperature–weight loss profile of a supported reagent can provide important quantitative information on the types of water present in the sample. It is usually possible to distinguish loosely held "physisorbed" water from strongly bonded "chemisorbed" water and helps determine the best conditions for removing the former. Thus, the TG of a bentonite clay (Fig. 3.7) shows distinct weight losses at ca. 144°C and ca. 536°C. The first differential of the weight loss plot ("DTG") allows the separation of the first (major) weight loss into two components. No further weight loss is observed above 600°C, although further high temperature thermal changes do occur and are visible by DSC (see Section 3.5.2).

The loss of water from supported reagents at temperatures greater than 300°C is probably due to surface dehydroxylation, which may be encouraged by the reagent, for example, in probable fluorodehydroxylation in supported fluorides.[15] TG can provide interesting comparative information on the residual water in supported reagents after employing different drying conditions. A comparison of the amount of "loosely bound" water (weight loss up to 300°C) against reagent loading for KF–alumina reagents dried under different conditions showed that at relatively low fluoride loadings (< 2 mmol/g), the reagents contain large amounts of

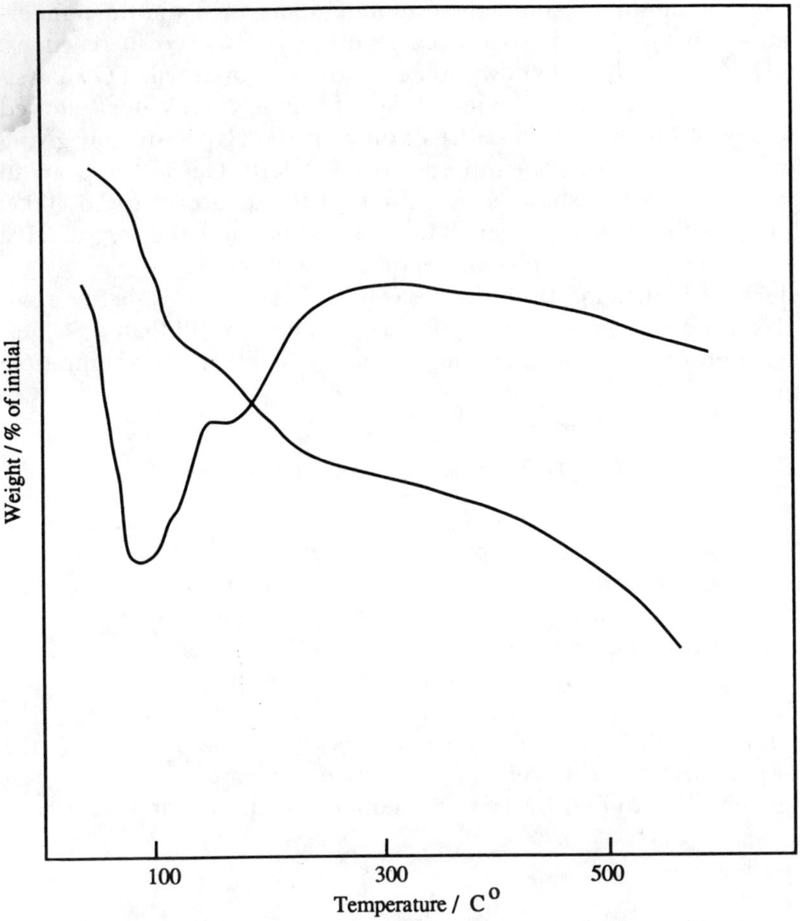

Figure 3.7 Thermogravimetric analysis (TG and DTG) of a bentonite clay showing a three-step loss of water.

water relative to KF and behave essentially as untreated alumina with regard to drying. At loadings of 2 mmole KF/g alumina and above the water content becomes almost directly dependent on the KF loading.[15] This suggests that the alumina surface is effectively covered by KF at the higher loadings so that any water present in these reagents (other than in the form of surface hydroxyls) is bonded to the reagent.

Loss or decomposition of the reagent can also be followed by TG. Most commercial instruments operate up to 700°C and may even go as high as 1500°C. This allows the observation of weight loss due to volatilization of even very stable, high boiling reagents.[15]

More usefully, the study of the decomposition of less stable reagents can help to throw light on the properties of some supported reagents and/

or help determine the maximum activation temperatures and the temperatures at which the supported reagents can be safely used in reactions. Clayfen (K10–$Fe(NO_3)_3$) is known to be an unstable material. TG analysis of clayfen is consistent with the initial loss of physisorbed water followed by the loss of N_2O_5, at a temperature of only ca. 100°C, presumably giving K10–Fe_2O_3. In contrast, the supported reagent K10–$CuCl_2$[27] appears to be more stable, but also shows a weight loss at temperatures up to 300°C in addition to that due to water. This may be because the loss of HCl causes the formation of copper oxides or oxychlorides.[26]

A calorimetric titration procedure has been shown to give better resolution of surface sites on carbon-supported palladium oxide than TG. This may be a useful method to study adsorption on solids that have multiple site character.[59]

3.5.2. Differential Scanning Calorimetry (DSC)

DSC (and DTA) can provide useful information on supported reagents where an exothermal or endothermal process is expected such as reaction between the support and the reagent, reaction of the material with air, or actual structural changes in the support.

Charcoal-supported metal ions are often good candidates for study by this technique as the charcoal can often reduce its supported ion. Thus, in the case of CuO–charcoal, the supported reagent was characterized by DSC in terms of the ability of the CuO to be chemically reduced by the charcoal support.[60] At different stages of aging, the amount of CuO was quantitatively determined by DSC by monitoring the thermal effects of the reaction:

$$4CuO + C \longrightarrow 2\ Cu_2O + CO_2$$

The CuO content can be determined from the measured heat change corresponding to the first DSC peak (Fig. 3.8). The heat evolved in the second peak corresponds to the reduction of Cu_2O to metallic Cu. As might be expected, the area of the first peak decreases with aging consistent with decreasing CuO content.[61] DSC can also be used to detect any further chemical changes over longer periods of aging that might result from reaction with the atmosphere such as the formation of carbonates from metal or metal-oxide-containing materials.

Structural changes in the supported reagent can be detected conveniently by DSC/DTA techniques. This can be especially useful for clay-based materials that can undergo such changes at elevated temperatures. Bentonite clays, for example, undergo an exothermal change to mullite at temperatures > 900°C. Such changes are, of course, undetectable by simple TG analysis.

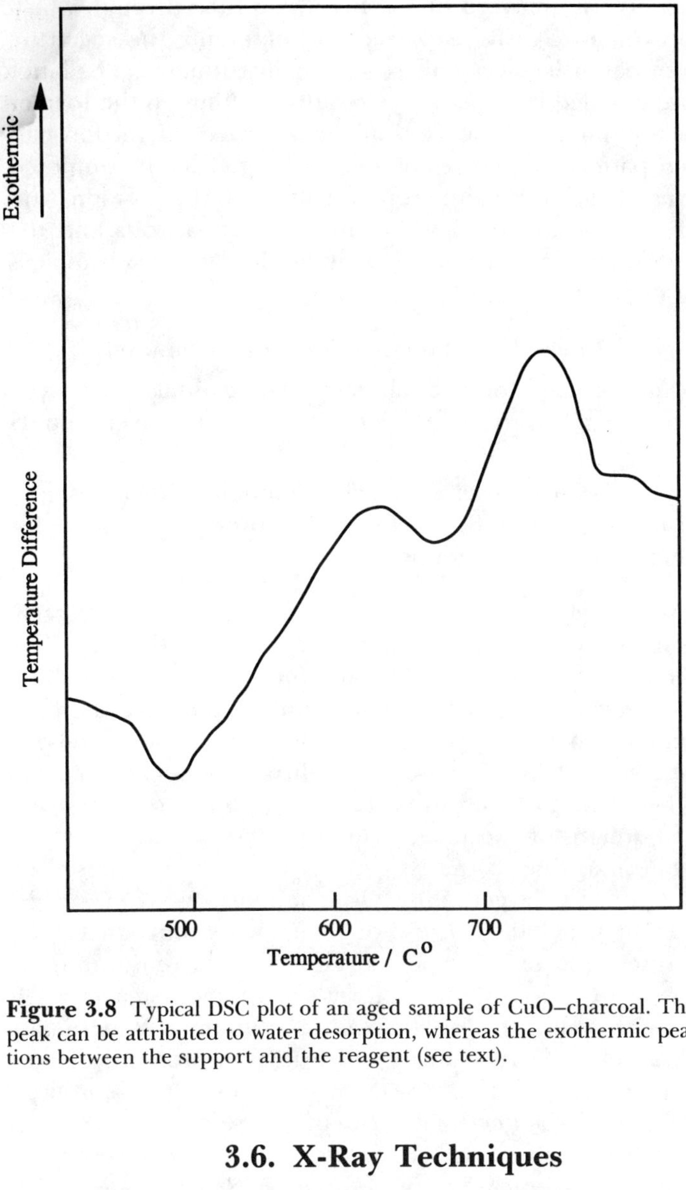

Figure 3.8 Typical DSC plot of an aged sample of CuO–charcoal. The first endothermic peak can be attributed to water desorption, whereas the exothermic peaks are due to reactions between the support and the reagent (see text).

3.6. X-Ray Techniques

3.6.1. Powder X-Ray Diffraction

Powder X-ray diffraction (XRD) is one of the most valuable techniques for the characterization of supported reagents. The technique can often be used to determine:

1. The dispersion of the reagent over the surface of the support
2. The identity of surface species
3. The basal spacing, d(001), in clay-based reagents

The powdered material is analyzed by placing it in a tube through which a beam of monochromatic X-rays is passed and observing the spectrum photographically or better by electronic recording instruments. The latter is required if more detailed information is required, although the former may be sufficient for simple comparison of the observed diffraction pattern with classified patterns. Mixtures of substances give a superimposed sum of the patterns of each crystalline component. Both the positions and the intensities of the line are used when making comparisons and the latter makes XRD capable of quantitative analysis. The major advantages of the diffraction method are:

1. Samples are analyzed directly (although a diluent may be used)
2. The analysis is quite straightforward, although the accumulation of data for dilute samples (e.g., supported reagents) may require several hours of exposure
3. Every crystalline substance gives a pattern, although amorphous substances cannot be detected and crystalline substances present at $< 1\%$ of the total sample weight may not be observable

Detectability limits probably represent the major drawback of the technique as applied to supported reagents, although in practice, they are good enough to measure dispersion efficiency and to detect "excess" reagent and other surface species present in significant amounts. Perhaps the most important advantage of the technique is its wide applicability.

At the simplest level, XRD can be used to confirm "reasonable" reagent dispersion over the surface of the support. For supported reagents containing ca. 10% by weight of the reagent (a fairly typical value), the absence of diffraction lines due to the reagent (typically superimposed on very diffuse lines due to the largely amorphous support—more crystalline supports may present problems because of their own strong diffraction pattern) or the presence of only very weak lines can be taken to indicate that most of the reagent is present in the form of extremely small crystals and/or is highly dispersed. Thus, in the case of sodium metaperiodate supported on silica gel (10% reagent by weight),[62] the presence of only very weak diffraction lines due to the reagent was interpreted as indicating, at worst, surface adsorbed crystals less than 50Å thick.[63]

XRD can be used similarly to detect "excess" poorly dispersed or macrocrystalline reagent. In the case of supported KCN, for example, reagents that showed significant excess cyanide by IR spectroscopy also showed X-ray lines corresponding to those of KCN.[64] In this case, the detection limit of KCN was shown to be as low as 0.2% of the total weight. More useful, supported reagents that are less amenable to spectroscopic analysis can be studied by XRD. In the case of supported fluorides, for example, XRD is probably the simplest and most direct technique to detect a poorly dispersed reagent.[15] In principle, XRD can be used to give quick information on the efficiency of dispersion of any supported reagent where the reagent normally exists in the crystalline state.

XRD may also be useful for the identification of species formed during the preparation on subsequent chemistry of a supported reagent. This can be especially useful for unstable reagents, corrosive reagents (i.e., those that are likely to attack the support), and where the supported reagent has been subject to high-temperature thermal treatment. XRD has been used to confirm the presence of K_3AlF_6 in some KF–alumina reagents.[65] Similarly, β-sodium aluminate was detected in γ-alumina, treated with sodium hydroxide and γ-alumina, which had been treated successively with sodium hydroxide then sodium metal.[66] Interestingly, the latter reagent did not show any diffraction lines due to the sodium itself. In contrast, α-alumina treated in the same way, while also showing the formation of β-sodium aluminate, did reveal the presence of sodium metal after successive treatment by NaOH and Na. The observed XRD differences between these materials helped to explain why the treated γ-alumina "superbase" proved to be the most active in olefin isomerizations. XRD is clearly the technique par excellence for such materials.

XRD (in conjunction with reflectance spectroscopy) has also been used to characterize the nature of surface species in $CuCl_2$ supported on aluminas and silica.[67] γ-Alumina samples, calcined at 400°C, for example, show $Cu_2(OH)_3Cl$. This species is absent in analogous silica materials.

The XRD measurement of the basal spacing, d(001), in expandable ("swelling") or pillared clays is of fundamental importance as it can be related to the interlayer spacing and hence the size of the trapped reagent or solvent and the space in which interlayer reactions can occur. Swellable clays include natural montmorillonites and synthetic materials such as saponite. These materials typically have silicate layer thicknesses of 9–10Å and the interlayer spacing in supported reagents based on these is simply calculated by subtracting this value from the measured basal spacing.

The interlayer spacing in clays is dependent on the molecules trapped between the silicate layers. Varying the solvent present can result in large variations in the interlayer spacing so that water-wet clays tend to give quite small d(001) values (although these are themselves dependent on the amount of water present[68–70]), whereas some aromatic solvents can give large values possibly due to their adopting a perpendicular orientation to the silicate layers so as to minimize electron–electron repulsion between negatively charged sites on the surface and the π electron cloud of the solvent. The presence of metal complexes can also result in large d(001) values. Thus, montmorillonite-supported copper(II) complexes, for example, show d(001) values as high as 20.7Å, which is an increase of 7.1Å compared with the "free" montmorillonite itself.[68–70] The largest values were observed for pyridine complexes of copper (II) and it has been suggested that this is also a result of the almost flat complex lying with its molecular plane perpendicular to the silicate sheets.

Basal region expansion as measured by XRD can be used as evidence of the successful incorporation of a reagent or other adsorbed molecule. Reaction of vacuum-dried sodium montmorillonte, $d_{100} = 9.4$Å, with chlorodiphenyl-phosphine caused an increase in the basal spacing to 14.5Å, which lends weight to the proposed reaction shown later.[71] Subsequent reaction with $PdCl_2(PhCH_2CN)_2$ resulted in a further increase in the measured basal spacing.

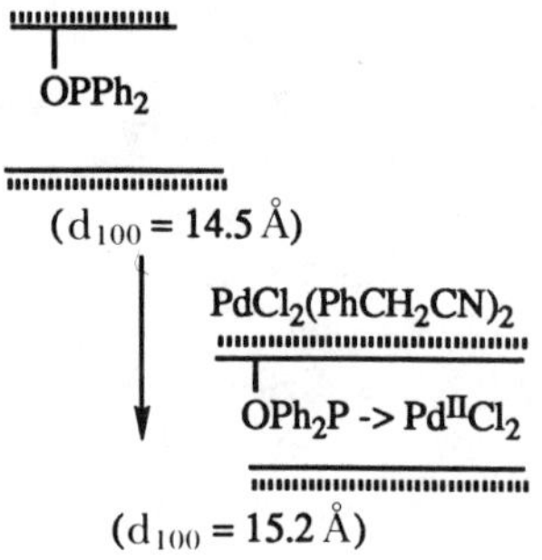

Pillared clays contain molecular props that hold apart the silicate layers and are generally designed to offer the possibility of shape selective catalysis while hopefully maintaining high thermal stability and other useful catalyst properties. XRD is an essential analytical tool for the characterization of such materials as the measurement of the interlayer spacing achieved by pillaring and its possible dependance on thermal and/or chemical treatment is of paramount importance. Basal spacings up to 20Å are quite common now and some of the materials have reasonable thermal stability.[72] The study of pillared clays is a subject of considerable interest and XRD is now established as a routine technique within this subject area.

3.6.2. Extended X-Ray Absorption Fine Structure (EXAFS)

X-ray absorption spectroscopy is a classic technique for the investigation of electronic structure and local environment of atoms in materials. Extended X-Ray absorption fine structure (EXAFS) is the name given to the oscillation on the high energy side of an X-ray absorption edge and is particularly useful to study the local environment around atoms in a variety of materials including amorphous solids.[73] Unfortunately the necessary equipment is rarely available outside of specialized surface science laboratories.

Surface EXAFS (SEXAFS) can be an excellent tool for surface structural determination as it is a direct probe of adsorbate atom, oxidation state, and environment,[74] and is particularly useful for very highly dispersed catalysts. For example, it has been applied to the study of supported cobalt–manganese catalysts prepared by so-called solvated metal atom dispersion.[75] The data obtained showed that the manganese was in the oxidized state, whereas the cobalt was in both metallic and oxidized states. A correlation of the SEXAFS results with activity of the materials showed that the most active catalysts for hydrogenation reactions had the largest amount of metallic cobalt.

The structure of the copper site in silica–copper(II) materials prepared in different ways has been studied by SEXAFS techniques.[76] This allowed the differentiation of mononulear and binuclear structures. The lowest dimensional structures with strong metal–metal interactions, as revealed by SEXAFS and related techniques, are thought to be the cause of the high catalytic activity of silica-supported copper(II) acetate.

3.7. Scanning Electron Microscopy

Scanning electron microscopy (SEM) can be used for the direct observation of the surface and particle profiles of supported reagents.[77] The technique effectively records differences in surface topology as the electron beam sweeps the sample. The magnification power of commercial electron microscopes allows easy observation of submicron-sized particles, therefore, SEM is a useful and fairly rapid technique to gain information on:

1. The dispersion of the reagent
2. The size and the shape of the particles of the reagent and the support

It should be noted, however, that it may prove difficult to observe very well-dispersed reagents of small particle size (although the failure to observe the reagent, as with XRD analysis, may be a useful indication of good dispersion) and that electron microscopy gives a picture of the profile of the particle and provides no direct information on its inner structure (unlike XRD, for example). Another important point to note is that most supported reagents are electrical insulators, therefore, sample charging, resulting from the accumulation of electrons on the material, may be a problem. This can result in distorted images although in practice the problem can be overcome by using low electron accelerating voltages and by coating the sample with a conducting material such as gold.

In spite of its simplicity, the technique has not been widely applied to the study of supported reagents, in part because of the unavailability of high magnification instruments in some laboratories and the difficulty in obtaining high resolution and informative images of irregular-shaped particles or highly amorphous materials. Analysis may also be largely restricted to the external surface, which in the case of many supported reagents represents only a small fraction of the area of interest.

The sophistication and value of the technique can be enhanced by adding an X-ray detector that can allow the determination of elements and their surface concentrations. This effectively allows the mapping of elements on the surface of materials and whereas its value as a technique is obvious it has rarely been applied to the analysis of the type of supported reagents on which this book focuses. It seems likely that this situation will change as commercial applications of supported reagents grow and with it the need for more precise information on the distribution of active sites.

SEM has been used successfully to study reagent dispersion and surface morphologies of alumina-supported metal fluorides, chlorides, and hydroxides.[78] MF–alumina materials prepared at or below the calculated loading for the ideal monolayer of MF showed images that were independent of M (M = Na, K, and Cs) with salt clusters of ca. 1 μm evident on the external surface. Significant differences between alumina-supported MF, MCl, and MOH were observed by SEM, which showed the effects of salt basicity and reagent surface affinity. The weakly basic chloride showed little affinity for the alumina surface and formed large clusters (10–50 μm), whereas the strongly basic hydroxide was well dispersed over the alumina surface but also caused severe surface erosion. Fluorides are in between these extremes and the SEM analysis may help us to understand the popularity of alumina–MF materials as basic reagents. The fluorides can be regarded as a compromise—offering reasonably high basicity without causing severe erosion of the support (at least at moderate activation temperatures and low loadings).

The use of SEM as a method to study reagent dispersion has been extended to the complex cyanide reagent $Na_2[Cu(CN)_3]$–alumina.[4] Micrographs obtained at different reagent loadings show that excess reagent appears in the form of well-defined crystals. The results agree nicely with those obtained from the use of FTIR to study reagent dispersion.[4] Similar studies have been carried out on sodium azide supported on zeolites. Loadings of NaN_3 on the Y-type zeolite CaY as high as 20 wt% gave micrographs very similar to the pure zeolite, whereas with higher loadings, large aggregates of NaN_3 were observed between microcrystals of the CaY.[79]

SEM can prove to be a more sensitive technique than XRD to study reagent dispersion.[80,81] Thus, in the case of CuO supported on activated carbon, SEM reveals the presence of unhomogeneously disturbed particles of 10–50 nm dimensions in samples that show no X-ray pattern due to CuO.[81]

3.8. Mass Spectrometry

Mass spectrometry (MS) has received little attention as a technique for the characterization of supported reagents. Although conventional electron impact MS has little obvious value in this area (except perhaps to observe evolved organic substrates or products) because of the low volatility of most reagents, secondary ion mass spectrometry (SIMS) is a more likely technique. A modification of SIMS is fast atom bombardment mass spectrometry (FABMS) where ions are sputtered from the unheated sample by high-energy neutral atoms. This technique has been used to help characterize supported reagents. It is possible to use FABMS to gain information on

1. Surface composition of the support and the effect of thermal or chemical treatment of the support
2. The identity of absorbed molecules
3. Reagent loading and dispersion

Both negative and positive ions can be detected by separate experiments but negative-ion FABMS tends to be much less sensitive than positive-ion FABMS.

The surface composition of zeolites has been studied by FABMS.[82] Acid treatment of ZSM-5, for example, has been shown to result in surface layers rich in silica. Interestingly, XRD and IR spectroscopy, unlike FABMS, did not reveal any differences in the surface composition on acid treatment. Similarly FABMS was used to show that high temperature steam treatment of H-ZSM-5 resulted in aluminium species being dislodged from the framework of the zeolite.[82]

FABMS has been used to identify the surface species present in various supported copper(I) reagents and especially those that are not amenable to study by more conventional techniques. Most of these materials clearly showed the presence of CuX_2^-. It is especially interesting to note that the highly reactive iodinating reagent KI–CuI–alumina showed I^- as the only significant species, whereas the much less reactive KI–CuI–charcoal showed the presence of complex anions.[83,84] The technique has been extended recently to the study of supported $CuSCF_3$.[84] A direct correlation of reagent loading with the intensity of Na^+ peak has been achieved for $Na_2[Cu(CN)_3]$–alumina.[4] This suggests that in some cases FABMS might be used to measure reagent loading. It should also be possible to study reagent dispersion by FABMS as excess crystalline reagent is more likely to produce high molecular weight cluster ions. Results should be treated with caution, however, in view of the low sensitivity of the technique, the inability to observe multiply charged ions, and the potential difficulty in distinguishing true surface species from those formed in the gas phase above the sample.

3.9. Nonspectroscopic Methods

This section deals with a number of useful nonspectroscopic measurements of surface characteristics of surface reagents. These are:

1. Acid and base strength and distribution of acid and base sites
2. Other specific functional sites (e.g., those that can be oxidized or reduced)
3. Surface area and other structural properties

Many of these parameters can be determined by spectroscopic methods and especially by "spectroscopic titration," but often alternative methods can be more convenient and obviously require less sophisticated equipment.

3.9.1. Acid and base strength and site distribution measurements

The classic method of acid strength measurements on solid catalysts is by noting the color changes of adsorbed dye indicators (Hammett indicators) during titration with a suitable base (typically a base) in a nonaqueous solvent.[85,86] A list of some suitable Hammett indicators for such measurements is given in Table 3.4. A simple "positive/negative" appraisal of the response of the solid to a range of indicators is sufficient to determine the maximum acid strength of a particular material.

Acidity can depend on the solvent. An aluminium-exchanged montmorillonite in a nonbasic solvent, such as toluene, was found to have acid sites up to $H_0 \leqslant -8.2$, whereas in a more basic solvent, such as an ether, the maximum acidity was reduced to $-5.6 < H_0 \leqslant -3.0$.[87] This can be attributed to the interaction of the ether with acid sites on the clay to reduce their acid strength.

In extreme cases even the most acidic of common indicators, 2,4-dinitrotoluene (pKa = −13.75) may be insufficient. This was found to be the case for some superacidic aluminas prepared by treating alumina with reagents including $AlCl_3$, P_2O_5, and PCl_5.[88]

Table 3.4 Indicators for Measuring Surface Acidity

Indicators	pKa
Neutral red	+ 6.8
4-(*p*-ethoxyphenylazo)-*m*-phenylenediamine	+ 5.0
4-phenylazo-1-naphthylamine	+ 4.0
4-*o*-tolylazo-*o*-tolusidine	+ 2.0
4-phenylazodiphenylamine	+ 1.5
2-nitrophenylamine	− 2.1
dicinnamalacetone	− 3.0
chalcone	− 5.6
anthraquinone	− 8.2
p-chloronitrobenzene	−12.7
m-chloronitrobenzene	−13.16
2,4-dinitrotoluene	−13.75

The acid strength distribution of various zeolites has been measured using the *n*-butylamine titration method and Hammett indicators.[89] For example, this can show how different degrees of cation exchange, the identity of the cation, and the aluminium content of the zeolite can affect zeolitic acidity.[90] It should be noted, however, that acidity measurements of small pore materials, such as zeolites, may depend more strongly on the size of the organic base than on the indicator molecules in that comparably smaller base molecules move inside the channels, whereas the larger indicator molecules remain in accessible sites and hence only indicate a limited number of sites. In practice this may simply slow down the titration (therefore, reaching the end point takes a long time—even up to several days) in which case ultrasonic activation may help or it may entirely preclude the use of the method for smaller pore materials.

Titrimetric determination of the amount of liberated base has been used to help study the nature of the basicity in some supported fluorides.[15,91] Washing with water and with aqueous HCl and titrating separately the solution gives measures of the "soluble base" and the "total basicity." It is interesting to note that in the case of the supported fluorides, there was no apparent correlation of soluble base that was very low, with observed catalytic activity that could be very high. On the other hand, the measured total basicity was extremely high and could not be accounted for by simple OH^- formation as shown in the equation:

$$12KF + Al_2O_3 + 3H_2O \longrightarrow 3K_3AlF_6 + 6KOH$$

These observations led the authors to suggest that although F^- alone hardly neutralizes aqueous HCl at all, F^-–Al_2O_3 is capable of acting as a strong base.[91]

3.9.2. *Measurement of Nonacidic or Basic Sites*

A fairly simple but very useful chemical probe of the efficiency of surface coating by a reagent is to measure the extent of adsorption of a second reagent. This is most likely to be effective when the first reagent prevents adsorption by the second reagent, as preferential adsorption by the supported reagent compared with free support surface is difficult to achieve. The adsorption of laurylamine was successfully used to probe the coverage of alumina surfaces by sodium cyanide.[92] Simple GLC analysis of a solution of the amine exposed to untreated alumina and to NaCN–alumina reagents showed a marked decrease in the amount of amine adsorbed onto the latter. No amine was adsorbed onto 5 mmol/g NaCN–alumina reagents suggesting complete insulation of the alumina surface by the cyanide.

The oxidation state of a metal in a supported reagent can be detected by redox titrations as can the amount of metal present in a particular batch of reagent. In this way the loading and oxidation state of iron in iron chloride–silica reagents has been determined.[93]

3.9.3. Measurement of Surface Area, Pore Volume, and Pore Diameter

One of the most important advantages of a supported reagent is the increase in effective surface area over the pure reagent. This increase in surface area is as much a function of the dispersion of the reagent as of the surface area of the support, but the measurement of the surface area of the supported reagent can still be informative as a reduction in value of the support to the supported reagent may be due either to (1) partial destruction of the structure of the support, or (2) the blocking of entrances to pores or surface region by microcrystals.

In practice, (1) is only likely to be a problem when the reagent is corrosive toward the support or when the activation step in the preparation involves high temperatures. Because most of the surface area of porous solids is contained within pores and monolayer dispersion of the reagent is rarely achieved (2) can be a serious problem. The use of small molecules, such as N_2 to measure surface area and other factors can also be a problem as this may not reveal sites inaccessible to larger molecules.

Surface areas are usually measured by the Brunauer-Emmett-Teller (BET) method, which relies on the adsorption of a monolayer of an inert gas (typically nitrogen or argon). Although this can be done manually based on a gas burette device, it is more common to use a commercial automatic surface area analyzer that can reduce the length of the experiment to less than 1 hr and can be made accurate for surface areas ranging from 10^3 m^2/g to < 1 m^2/g. In principle, at least, it may be possible to modify the experiment to allow specific adsorption onto just reagent or free surface sites, but it would be very difficult to select such a highly specific adsorbent that was also amenable to the type of experiment. Spectroscopic titration is probably a more useful technique to probe the surface in such detail.

Reagents that are likely to be corrosive toward a support material are those that have a high affinity for the structural atoms of the support. Thus, OH^- and F^-, in particular, are likely to attack silicon and aluminium centers causing destruction of the support. The measured surface areas of supported metal fluorides dried at different temperatures and supported metal chlorides have been compared over a range of reagent loadings.[15]

The following conclusions were made:

1. An increase in reagent loading generally causes a decrease in the surface area of the supported reagent consistent with the blocking of pores and surface irregularities

2. The more corrosive fluorides cause a much more rapid decrease in surface area with increased loading of reagents

3. High temperature drying of the supported reagent causes a large reduction in the surface area

Surface areas of less than 20 m^2/g were measured for high temperature (600°C) dried KF–alumina samples based on an alumina surface area greater than 120 m^2/g. This represents a dramatic destructuring of the support consistent with the formation of K_3AlF_6.[15]

Even when surface corrosion is unlikely to be a significant problem, the presence of microcrystals on the support can cause a reduction in effective surface area. Thus the deposition of $K_2Cr_2O_7$ onto various high surface area and low surface area supports has been found to result in a significant reduction in the measured surface area of the supported reagent.[94] In the case of a silica with a surface area of 395 m^2/g, the supported reagent $K_2Cr_2O_7$–silica (4% by weight chromium) had a surface area of only 254 m^2/g. In the case of the lower surface area support BPO_4, the corresponding reduction was from 11 to <4 m^2/g. The presence of dichromate microcrystals on the supports surface was confirmed by X-ray diffraction.

The measurement of surface areas can also be used to screen possible support materials. The most effective KF–CaF_2 reagents for organofluorination reactions were found to be based on the highest surface area (precipitated) calcium fluorides.[95] The resulting reagent has a much higher surface area (14 m^2/g) than that of natural fluoride (unmeasurable value) or potassium fluoride itself (0.1 m^2/g).

Pore volume and average pore diameter can also be measured by the BET method.[23,94] Both of these values might reasonably be expected to drop in value compared with the free supports if the reagent enters the pores and such reductions might be used to determine the extent of adsorption on the internal as opposed to the external surface of the support.

REFERENCES

1. T. J. Pinnavia, R. Raythatha, J. G-S. Lee, L. J. Halloran and J. F. Hoffman, J. Am. Chem, Soc., (1979), **101,** 6891.
2. M. Onaka, K. Sugita and Y. Izumi, J. Org. Chem., (1989), **54,** 1116.
3. T. Ando, J. H. Clark, D. G. Cork, M. Fujita and T. Kimura, J. Org. Chem., (1987), **52,** 681.
4. C. V. A. Duke, J. M. Miller, J. A. Clark and A. P. Kybett, Spectrochimica Acts, (1988), **44A,** 1207.
5. R. K. Vincent and G. R. Hunt, Aypl. Optics, (1968), I, 53.
6. S. L. Regen, S. Quici and M. D. Ryan, J. Am. Chem. Soc., (1979), **101,** 7629.
7. H. C. Foley, S. J. De Canio, K. D. Tan, K. J. Chao, J. H. Onuferko, C. Dybowski and B. C. Gortes, J. Am. Chem. Soc., (1983), **105,** 3074.
8. N. Kakuta, A. Kazusaka, A. Jamazaki and K. Miyakora, J. Chem. Soc. Farad I, (1984), **80,** 3245.

9. A. Al-Jazzaa, J. H. Clark and M. Robertson, Chem. Lett., (1982), 405.
10. M. D. Ward, T. V. Harris and J. Schwartsz, J. Chem. Soc. Chem. Commun., (1980), 357.
11. J. H. Clark, C. V. A. Duke, J. M. Miller and S. J. Brown, J. Chem. Soc. Chem. Commun., (1986), 877.
12. J. H. Clark and C. V. A. Duke, J. Org. Chem., (1985), **50,** 1330.
13. J. H. Clark, C. V. A. Duke, S. J. Brown and J. M. Miller, Spectrochimica Acta, (1986), **42A,** 811.
14. J. H. Clark, C. W. Jones, C. V. A. Duke and J. M. Miller, J. Chem. Soc. Chem. Commun., (1989).
15. T. Ando, S. J. Brown, J. H. Clark, D. G. Cork, T. Harafusa, J. Ichihara, J. M. Miller and M. S. Robertson, J. Chem. Soc. Perkin II, (1986), 1133.
16. L. M. Weinstoch, J. M. Stevenson, S. S. Tomellini, S. H., Pan, T. Utne, R. B. Jobson and D. F. Reinhold, Tetrahed. Lett., (1986), 3845.
17. J. H. Clark and A. P. Kybett, unpublished results.
18. J. H. Clark, C. V. A. Duke, A. P. Kybett and J. M. Miller, Spectrochimica Acta, (1990), **46A,** 1381.
19. D. T. B. Tennatroon, W. Jones, J. M. Thomas, J. H. Ballantine and J. H. Purnell, Solid State Ionics, (1987), 205.
20. J. H. Clark and C. V. Duke, J. Org. Chem., (1985), **50,** 1330.
21. R. S. Drago and E. E. Getty, J. Am. Chem. Soc., (1988), **110,** 3341.
22. A. Krzywicki and M. Marczewski, J. Chem. Soc., Farad. I, (1980), **76,** 1311.
23. T. Ishikawa, E. Amano, M. Muroya and S. Kondo, Chem. Lett., (1983), 415.
24. M. A. Kohler, N. W. Cant, M. S. Wainwright and D. L. Trimm, J. Catal., (1989), **117,** 188.
25. A. F. J. Wielers, A. J. H. M. Kock, C. E. C. A. Hgo, J. W. Gens and A. M. Van der Kraan, J. Catal. (1989), **117.**
26. O. Radhwi, MSc. Thesis, University of York, England, (1989).
27. J. H. Clark, A. P. Kybett, K. J. Macquarrie, S. J. Barlow and P. Landon, J. Chem. Soc. Chem. Commun., (1989), 1353.
28. N. Davison and W. R. McWhinnie, Inorganica Chimica Acta, (1987), **131,** 9.
29. M. S. Climent, J. M. Marinas and J. V. Sinisterra, React. Kinet. Catal. Lett., (1986), **32,** 177.
30. C. G. Flowers, S. Lindley and J. E. Leffler, Tetrated, Lett., (1984), 499.
31. J. H. Anderson, J. Catal., (1972), **26,** 277.
32. D. Briggs, Surf. Sci., (1987), **189–190,** 801.
33. J. E. Fulghum, G. E. McGuire, I. H. Musselman, R. J. Nemanich, J. M. White, D. R. Chopra and A. R. Chourasia, Anal. Chem., (1989), **61,** 243.
34. N. Kututa, A. Kazusaka and K. Miyahara, Bull, Chem. Soc. Jpn., (1986), **59,** 3267.
35. R. B. Friedman, J. J. Freeman and F. W. Lytle, J. Catal., (1978), **55,** 10.
36. R. B. Shalvoy, P. J. Reucroft and B. H. Davis, J. Catal., (1979), **56,** 336.
37. R. Eckman, J. Phys. Chem., (1982), **76,** 2767.
38. A. J. Vega, Z. Luz, J. Phys. Chem., (1987), **91,** 365.
39. Z. Luz and A. J. Vega, J. Phys. Chem., (1987), **91,** 374.

40. C. E. Bronnimann, I. S. Chuang, B. L. Hawkins and G. E. Macid, J. Am. Chem. Soc., (1987), **109,** 1562.
41. A. J. Vega. J. Am. Chem. Soc., (1988), **110,** 1049.
42. F. Freude and J. Klinowski, J. Chem. Soc. Chem. Commun., (1988), 141.
43. M. W. Anderson and J. Klirowski, J. Am. Chem. Soc., (1990), **112,** 10.
44. J. H. Clark, E. M. Goodman, D. K. Smith, S. J. Brown and J. M. Miller, J. Chem. Soc. Chem. Commun., (1986), 657.
45. G. E. Balliman, C. J. Groombridge, R. K. Harris, K. J. Packer, B. J. Say and S. F. Tanner, Philos. Trans. R. Soc. London, Ser. A, (1981), 299, 643.
46. A. Naito, S. Ganapathy, J. Akasaka and C. A. McDowell, J. Chem. Phys., (1981), **74,** 3(190).
47. A. Naito, S. Ganapathy and C. A. McDowell, J. Phys. Chem., (1981), **74,** 5393.
48. N. Zumbulyadis, P. M. Henrichs and R. H. Young, J. Phys. Chem., (1981), **75,** 1603.
49. A. Naito, S. Ganapathy and C. A. McDowell, J. Mag. Reson., (1982), **48,** 367.
50. L. M. Gierasch, M. H. Frey, J. G. Hexem, S. J. Opella, "New Methods and Applications of NMR Spectroscopy," G. D. Levy, Ed., Am. Chem. Soc., Washington DC, (1982), ACS Symp. Ser. 191, p. 233.
51. J. G. Hexem, M. H. Frey and S. J. Opella, J. Phys. Chem., (1982), **77,** 3847.
52. J. M. Serratosa, J. A. Rousell-Colom and J. Sanz, J. Mol. Cat., (1984), 225.
53. L. B. Ebert, P. A. Huggins and J. I. Brauman, J. Chem. Soc. Chem. Commun., (1974), 924.
54. J. Karger, H. Pfeifer, S. Rudtsch and W. Heink, J. Fluorine Chem., (1988), 349.
55. H. Selig, M. Robinovit, I. Agranat, C. H. Li and L. Ebert, J. Am. Chem. Soc., (1976), **98,** 1601.
56. D. R. Brown and L. Kevan, J. Phys. Chem., (1988), **92,** 1971.
57. R. K. Kukkagapu and L. Kevar, J. Phys. Chem., (1988), **92,** 607.
58. O. Braddell, R. C. Barklie, D. H. Doff, N. H. J. Cangas and A. McKimm, Z. Phys. Chemie, (1987), **151,** 157.
59. Y. Y. Lin, R. S. Drago, M. W. Babrich, N. Wang and P. E. Dean. J. Am. Chem. Soc., (1987) **109,** 169.
60. N. Kakuta, A. Kazusaka and K. Miyakara, Bull. Chem. Soc. Jpn., (1986), **59,** 3267.
61. P. Ehrburger, J. M. Henlin and J. Lahaye, J. Catal., (1986), **100,** 429.
62. D. N. Gupta, P. Hodge and J. E. Davies, J. Chem. Soc. Perkin 1, (1981), 2970.
63. J. M. Bijvoet, N. H. Kolkmeyer and CH. Macgillavry, "X-Ray Analysis of Crystals," Butterworths, London, (1951), 51.
64. J. H. Clark and C. V. A. Duke, J. Org. Chem., (1985), **50,** 1330.
65. T. Ando, J. H. Clark, D. G. Cork, T. Harafusa, J. Ichihara and T. Kimura, Tetrahed. Lett., (1987), 1421.
66. G. Suzukamo, M. Fukao and M. Minobe, Chem. Lett., (1987), 585.
67. C. Zipelli, J. C. J. Bart, G. Petrini, S. Gawagno and C. Cimino, Z. anorg. allg. Chem., (1983), **502,** 199.

68. W. F. Bardley, G. F. Clark and R. E. Grim, Z. Krist., (1937), **97,** 216.
R. W. Mooney, A. G. Keenan and L. A. Wood, J. Am. Chem. Soc., (1952), **74,** 1367.
69. K. Norrish, Nature (London), (1952), **173,** 256.
70. N. Davison and W. R. McWhinnie, Inorganica Chim. Acta, (1987), **131,** 9.
71. B. M. Chondary, K. R. Kuma, Z. Jamil and G. Thyagarajan, J. Chem. Soc. Chem. Commun., (1985), 931.
72. K. Urabe, H. Sakurai and Y. Izumi, J. Chem. Soc. Chem. Commun., (1986), 1074.
73. D. C. Koninsburg and R. Prims, "X-Ray Absorption: Principles, Applications, Techniques of EXAFS, SEXAES and XANES," Wiley, N.Y., (1988).
74. J. E. Fulghum, G. E. M. McGuire, I. H. Musselman, R. J. Nemanich, J. M. White, D. R. Chapra and A. R. Chowasia, Anal. Chem., (1989), **61,** 243.
75. B. J. Ian, K. J. Klabunde, T. Tanaka, H. Kanai and S. Yaskida, J. Am. Chem. Soc., (1988), **110,** 5951.
76. M. Nomura, A. Kazusaka, N. Kakuta, Y. Ukisu and K. Miyahara, J. Chem. Soc. Farad. Trans. I, (1987), **83,** 1227.
77. Y. Zhang, Inorg. Chem., (1982), **21,** 3889.
78. T. Ando, J. H. Clark, D. G. Cork and T. Kimura, Bull. Chem. Soc. Jpn. (1986), **59,** 3281.
79. M. Onaka, K. Sugita and Y. Izumi, Chem. Lett., (1986), 1327.
80. T. Ando, J. H. Clark, D. G. Cork and T. Kimura, Bull. Chem. Soc. Jpn., (1986), **59,** 3281.
81. P. Ehrburger, J. M. Henlin and J. Lahaye, J. Catal., (1986), **100,** 429.
82. J. Dwyer, F. R. Fitch, G. Gin and J. C. Vickerman, J. Phys. Chem., (1982), **86,** 4574.
83. J. H. Clark, C. W. Jones, C. V. A. Duke and J. M. Miller, J. Chem. Res.(S), (1989), 238.
84. J. H. Clark, C. W. Jones, A. P. Kybett, M. A. McClinton, J. M. Miller, K. Bishop and R. J. Blade, J. Fluorine Chem., (1990), **48,** 249.
85. Ch. Walling, J. Am. Chem. Soc., (1950), **72,** 1164.
86. H. A. Benesi, J. Am. Chem. Soc., (1956), **78,** 5490.
87 M. Kawai, M. Onaka and J. Izumi, Bull. Chem. Soc. Jpn., (1988), **61,** 1237.
88. A. Krzywicki and M. Marczewski, J. Chem. Soc. Farad. I, (1980), **76,** 1311.
89. A. Corma, H. Garcia and J. Primo, J. Chem. Res.(S), (1988), 40.
90. W. F. Kladnig, J. Phys. Chem., (1979), **83,** 765.
91. T. Ando, J. H. Clark, D. G. Cork, T. Harafusa, J. Ichihara and T. Kimura, Tetrahed. Lett., (1987), 1421.
92. S. L. Regen, S. Guici and M. D. Ryan, J. Am. Chem. Soc., (1979), 7629.
93. T. C. Jempty, K. A. Z. Gogins, Y. Mazur and L. L. Miller, J. Org. Chem., (1981), **46,** 4545.
94. M. S. Climent, J. M. Marinas and J. V. Sinisterra, Tetrahedron, (1988), 2963.
95. J. Ichihara, T. Matsuo, T. Hanafusa and T. Ando, J. Chem. Soc. Chem. Commun., (1986), 793.

4 Case Studies

4.1. Supported Fluorides

The versatility of the fluoride ion in synthetic chemistry[1]—in organofluorinations, base-catalysis, the removal of silyl protecting groups, and polymerization—has led at least in part to the enormous research effort in the study of supported fluorides. In relation to the entire field of supported reagents, however, this effort seems disproportionate and is largely due to their apparently erratic behavior in reaction (ranging from highly reactive to totally unreactive in organofluorinations, for example), the controversy over their exact nature and composition, and the progress that has been made with their application in at least a semitechnical capacity.

The most significant early work on supported fluorides is due to Ando and co-workers[2,3] who reported the use of KF–celite and KF–alumina as useful heterogeneous reagents for alkylation reactions. Their research included the use of these materials in the synthesis of crown ethers[3] and in the *N*-alkylation of amides and *N*-heterocycles[4]—reactions that normally require strong bases. Ando et al.[5] quickly appreciated the value of supported reagent optimization and after early investigations on the methods of preparing KF–alumina, they published, in 1982, the classic paper on optimization of the efficiency of supported fluorides. The effects of the support (reported earlier[3]), the metal cation, the loading, and the reaction solvent were considered. In alkylation reactions, the order of effectiveness of the supports was found to be: alumina > molecular sieves > celite. Reagents with loadings of 1, 2, 5, and 10 mmole MF/g alumina were prepared using LiF, NaF, KF, and CsF along with tetra-*n*-butylammonium floride (TBAF). Comparison with unsupported fluorides showed that the supported versions were generally more reactive and that in both cases the effect of changing M was CsF>KF>>NaF>LiF. Similarly TBAF–alumina, as with TBAF itself, was found to be especially reactive, but Ando et al. concluded that the difficulty in handling the highly hygroscopic TBAF reagents precluded their use as convenient reagents. Increased loading of MF was also found to result in increased reactivity. More polar solvents were found to improve the effectiveness of supported fluorides in alkylation reactions but perhaps more significantly, it was noted that reactions could be achieved even in hydrocarbon solvents.

The early work of Ando et al. was limited to non catalytic reactions and these require the use of large amounts of material (typically 2.5 mEquiv. of MF with respect to substrate). The work, however, did provide important information on reagent optimization and highlight the practical advantages of the use of supported reagents.

Probably the main use of supported fluorides has been to catalyze Michael reactions. KF–alumina has been shown to be a very effective catalyst for these important carbon–carbon bond-forming reactions and its reactivity is considerably greater than many other fluoride ion sources, including KF–18–crown–6.[6] Forcing drying condition (e.g., 150°C for 24 hr under reduced pressure) to remove the bulk of physisorbed water (conditions determined by the use of thermogravimetry) and quite high loading (ca 2 mmol/g) were found to give the best results. In contrast to the earlier studies on relative reactivities using the stoichiometric alkylation reactions as models, higher fluoride loadings resulted in a fall in reactivity in these *catalytic* reactions.[6]

Significant improvements in rate or efficiency of many other Michael reactions using KF–alumina have been reported.[6–11] The reagent was found to be capable of undergoing 150+ turnovers with little loss in efficiency,[7] could generally enable reactions to be run at room temperature, and, if desired in the absence of solvent.[11] KF–alumina has been reported to be superior to many other catalysts for Michael reactions including powerful bases such as sodium methoxide (it has the additional advantage of being very resistant to the presence of water)[9] and in many respects is an excellent catalyst for such reactions.

In addition to Michael reactions KF–alumina has been used successfully in a variety of base-catalyzed organic reactions, including Wittig, Wittig–Horner, and Knoevenagel reactions,[12] the synthesis of phosphoric esters,[13] α-nitroalcohols,[14] 1,4-diketones,[8] and pseudoionones.[15] In many cases the reactions were carried out in the absence of solvent, although perhaps the most significant reactions were carried out by percolating the substrate or substrates dissolved in methylene chloride through a bed of KF–alumina.[16] These "chromatography reactions" gave good yields of generally very pure products and, at least for catalytic C-C bond forming condensations, the KF–alumina was reusable after washing and thermal activation. Flash chromatography was found to be especially effective for larger-scale reactions.[16]

The removal of at least most of the physisorbed water from KF–alumina at temperatures of 100–200°C in vacuo does seem to be advantageous to its performance as a base. The optimum loading is still far from clear, however, with loadings from 1 to 20 mmol/g and even higher having been reported to be effective. The low cost of the material, its stability toward heat and water, and its ease of handling make it a very attractive basic catalyst that may well find industrial applications in the near future.

KF enjoys a widespread use as a source of nucleophilic fluorine, therefore, it is perhaps surprising that in research and chemical manufacturing it should, when supported, be used mostly as a basic reagent or catalyst. In fact, the residual fluorine nucleophilicity of alumina-supported KF is minimal because of a strong surface OH . . .F^- hydrogen bonding.[17] With this in mind, it is possible to design supported fluorides that do not suffer from this kind of interaction. Perhaps the most famous example is KF–CaF_2, which was discovered and reported independently by groups in the United Kingdom[17] and Japan.[18] Calcium fluroide, which is itself ineffective in nucleophilic fluorinations, can be prepared as a reasonably high surface area material (ca. 14 m^2/g)[18] with an effectively inert surface. The potassium fluoride can then be loaded onto this support by the conventional evaporation technique using a methanolic solution of KF slurried with the CaF_2[17] or simply by thorough mixing.[18] Although the maximum surface area of KF that can be achieved in this way (14 m^2/g) is much less than that of more traditional supports (>100 m^2/g), it is much greater than that of ordinary KF (0.1 m^2/g).[18] KF–CaF_2 is certainly superior to KF for replacing heavier halogens by fluorine, at least with aliphatic substrates.[17,18] Both high-resolution solid-state ^{19}F NMR[17] and XRD[18] show that unlike KF–alumina, KF–CaF_2 has no apparent interaction between the reagent and the support.

Using the same logic, organic polymers, which also can be prepared with no surface hydroxyl groups, can be used to support KF for use in organofluorinations.[1] High-surface area (>600 m^2/g) cross-linked hydrocarbon polymer–KF reagents have been successfully used to convert acid chlorides to acid fluorides in high yields.[19]

Supported fluorides are among the most popular supported reagents and the most difficult to characterize. There is no single technique that can be used to give useful information about the materials. As a result, a multitude of techniques, including SEM, NMR, IR, XRD, and FABMS, have been used to obtain information about reagent dispersion, nature of the adsorbed species, and reagent–support interactions.

SEM analysis has shown that metal fluorides are not as corrosive as hydroxides or as neutral as chlorides toward alumina.[20] The KF forms clusters of about 1 μm diameter in low loaded (2 m^2/g), moderately dried (90°C under vacuum) KF–alumina reagents. High-resolution solid-state ^{19}F NMR of various supported fluorides gives evidence of both $M^+ \ldots F^-$ interactions and $F^- \ldots$ support interactions.[21] It is interesting to note that the differences in chemical shift from pure fluoride to MF–alumina is the same for different M^+.

In the most comprehensive paper published on the characterization of alumina-supported fluoride reagents, a series of analytical techniques were used to optimize the reagent preparation and learn about the active species in the reagents.[22] Perhaps the most important conclusion of this work is that a combination of low MF loadings and highly efficient drying procedures give the best reagents for catalytic reactions, whereas somewhat higher loadings and more moderate drying procedures give the best reagents for non catalytic reactions. These observations are consistent with the earlier work on alkylations and Michael reactions. The identification of the active species on the supported fluorides proved to be less straightforward, although the authors argued that the formation of surface OH^- cannot be entirely responsible for the observed high basicity of the reagents, as shown in the equation:

$$12KF + 3H_20 + Al_2O_3 \longrightarrow 6KOH + 2K_3AlF_6$$

Coordinatively unsaturated surface-based F^- ions were also thought to play an important role.[22] A rather complex schematic model involving surface OH groups and F^- ions was proposed.[22] It was also noted that the combination of high loadings and high drying temperatures resulted in significant damage to the alumina structure with resulting loss in activity.

In a later paper, Weinstock and co-workers[23] argued that the high reactivity of KF–alumina compared with other fluoride-containing basic systems is in fact due to the reaction of the F^- ion with the alumina to produce potassium hexafluoroaluminate, potassium hydroxide, and potassium aluminate. Their evidence for this reaction was based on IR and Raman analysis (which showed bands due to the AlF_6^{3-} ion as well as K_2CO_3) and on the observation that KOH–alumina reagents, of comparable reactivity in the reaction of chloroform with *m*-nitrobenzaldehyde, could be prepared. In a response to this, Ando and co-workers[24] gave further evidence to suggest that the reactivity of KF–alumina cannot be simply associated with surface OH^-. Their evidence included a titrimetic analysis of NaF–alumina and KF–alumina basicities clearly showing that the amount of water-soluble base does not correspond to the observed catalytic activities and that the basicity of F^- dramatically increased when it was supported on alumina. They concluded that the efficiency and versatility of KF–alumina as a heterogeneous base for organic synthesis is due to at least three possible mechanisms:

1. Dispersion and increased surface area of KF giving coordinatively unsaturated F^-
2. Liberation of strong base during preparation
3. The cooperative action of F^- and the hydrated alumina surface

It seems likely that although KF–alumina continues to enjoy widespread application as a base in organic synthesis, the controversy over its exact nature continues. Weinstock's observation of carbonate may prove to be especially important as it has been suggested recently that this arises from the formation and subsequent reaction with atmospheric CO_2 of surface potassium oxide, which would itself act as a possible base. If this is the case, then the activity of KF–alumina could prove to be even more critically dependent on the method of preparation than has been assumed, because K_2O is likely to be quickly destroyed by exposure to the atmosphere.

4.2. Supported Thiocyanates

Organic thiocyanates are important intermediates in heterocyclic synthesis.[25] These can be prepared from salts containing the thiocyanate ion. Supported reagents have been used recently to activate, for example, KSCN and CuSCN. These procedures have increased the ease and the scope of reactions involving the thiocyanate ion.

The following section illustrates the reaction of these supported reagents, and how improvements in their reactivity have resulted from the use of relatively simple analytical techniques.

4.2.1. KSCN–SiO_2

The first report of KSCN–SiO_2 was by Brimeyer et al.[26] They prepared the reagent using two methods, grinding KSCN and silica together, and an evaporative technique. Both reagents are equally efficient in the reactions detailed in this paper.

The reactions were not preparations of organic thiocyanates, but were conversions of epoxides to episulfides:

These reactions were carried out by stirring an excess of the supported reagent and the epoxide in toluene, followed by filtration and evaporation of the solution.

The supported reagents had a loading of 10 mmol/g and were used in a 10-fold excess of KSCN to substrate. A small amount of water was found to be necessary for optimum activity but no details were given. The yields

$$PhCH_2Cl \xrightarrow[\text{30°C—4 h, no solvent}]{\text{KSCN/support}} PhCH_2SCN$$

Support	Yield (based on $PhCH_2Cl$) (%)
SiO_2	95
Al_2O_3	66
MS5A powder	13
MS13X	7
Kieselguhr powder	4

Figure 4.1 Effect of support on activity of supported KSCN.

were moderate to good based on substrate, but less impressive based on supported reagent.

The evaporative reagent was also used by Kodomari et al.[27] in the preparation of organic thiocyanates. Again the loading was 10 mmol/g. Other supports were used and silica was found to be the best, with neutral alumina also good (Fig. 4.1):
Again, the supported reagents were used in excess and the best yield based on KSCN was ~15%

Various alkyl and benzyl halides reacted to give the corresponding thiocyanates under the same conditions. Thus, the KSCN–SiO_2 reagent can be used in a mild and convenient manner although it does suffer from the drawback of inefficient use of KSCN.

4.2.2. Optimization of Reagents

Loadings as high as 10 mmol/g are now recognized to be too high even for high surface area supports and it is normally true that 0.5–2 mmol/g is the optimum loading range. It was found that by studying the reagent preparation, significant improvements could be made to these reagents.

Ando et al.[28] studied the infrared spectra of some of these reagents, and used results to improve the activity of the supported reagents. The SCN^- anion has a strong absorption around 2000 cm^{-1}, conveniently situated away from the broad strong support absorptions at lower frequencies, and also away from the broad OH bands associated with most supports. Thus, this supported reagent can be easily studied by infrared spectroscopy, in this case using KBr discs as a sampling technique.

The results for KSCN–SiO_2 at 10 mmol/g and 3 mmol/g are shown in Figure 4.2. The spectra show a maximum for KSCN (unsupported) at 2050 cm^{-1}. In the 3 mmol/g reagent this band has disappeared, to be replaced by a band at 2070 cm^{-1}. This shift is due to a significant reduc-

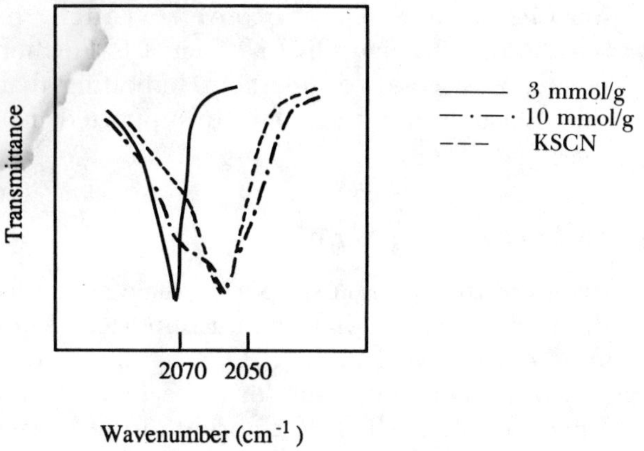

Figure 4.2 Effect of loading on $v(SCN^-)$.

tion in ion-pairing, indicating a well-dispersed layer of KSCN in the 3 mmol/g reagent. The 10 mmol/g reagent shows both bands, but predominantly the unsupported KSCN band, indicating that the KSCN is not well dispersed, most of it being present in large clusters. This proves that the reagent is substantially overloaded.

KSCN–Al_2O_3 gives similar spectra, but with a lower surface area, and are actually somewhat worse than the SiO_2 reagent. Reactions of these supported reagents were also studied, and the optimized reagents were found to be superior to the overloaded reagents in efficiency (with respect of SCN^-) and also generally gave faster rates. CaF_2, K10, and florisil were also shown to be useful supports when the loadings were optimized.

Some results are shown in Figure 4.3. It is not clear whether these results reflect accurately the efficiencies of the reagents. To accomplish this excess $PhCH_2Cl$ is needed. Runs 1–4 (Fig. 4.3) show the effect of loading on silica. With no support, a very slow reaction occurs. A 2 mmol/

Figure 4.3 Effect of loading on reactions of KSCN–support.

$PhCH_2Cl$ + KSCN/support ⟶ $PhCH_2SCN$
1 mol : 2.5 mol KSCN

Run	Support	Loading (mmol/g)	Time (h)	Conversion (%) based on $PhCH_2Cl$
1.	none	—	5	25
2.	SiO_2	2	3	62
3.	SiO_2	3	1	96
4.	SiO_2	10	3	96
5.	Al_2O_3	0.9	0.5,3	45,23
6.	Al_2O_3	2	0.5,3	92,77
7.	Al_2O_3	10	0.5,3	32,68

g reagent gives a much faster rate, leading to a 62% conversion after 3 h. A 3 mmol/g loading increases the rate dramatically to 96% in 1 h. Further increasing to 10 mmol/g slows the reaction considerably, indicating that the overloaded reagent is less reactive than the correctly supported material.

4.2.3. *Effect of Exposed Surface*

A comparison of the alumina-supported reagent results shows a similar pattern. All are more active than the unsupported material, with a definite optimum activity at ~2 mmol/g. One significant difference, however, is apparent with the alumina-supported reagent. Run 5 (Fig. 4.3) shows a rapid conversion of the benzyl chloride to the benzyl thiocyanate product using the 0.9 mmol/g reagent. As the reaction proceeds, the conversion drops significantly from 45% in 0.5 h to 23% in 3 h. A similar, although less pronounced, effect is also found with the 2 mmol/g reagent (92% in 0.5 h to 77% in 3 h). At the longer times dibenzyl disulfide is formed, and this decreases the conversion. Dibenzyl disulfide is formed from benzyl thiocyanate by a base-promoted reaction and thus the conversion to benzyl thiocyanate decreases with time. This reaction appears to be promoted by the alumina surface, which contains basic sites, and therefore, is less of a problem with the 2-mmol/g reagent than with the 0.9 mmol/g reagent, which presumably has more exposed surface (a fact that is backed up by the stronger adsorption of organics onto the surface of the lower loaded supports). Silica-supported reagents do not suffer from this problem because silica does not have basic sites. The 10 mmol/g Al_2O_3 reagent, where the Al_2O_3 surface is completely hidden, does not promote this side reaction, and this is, in some respects, superior to the more reactive 2 mmol/g reagents.

4.2.4. *Allyl and t-*Alkyl Substrates Isomerization

These reagents have also been used in the formation of allyl thiocyanates. In these reactions, the formation of the more stable isothiocyanate product occurs to a greater extent:

$$PhCH{=}CHCH_2X \xrightarrow{KSCN/Y} PhCH{=}CHCH_2SCN + PhCH{=}CHCH_2NCS$$

$X = Br, Cl$ $\quad Y = Al_2O_3$, SiO_2, or CaF_2

Again the optimum reagent, with respect to reaction rate, was the 3 mmol/g SiO_2-supported reagent. On prolonged stirring, however, isomerization

X	Y	Loading (mmol/g)	Time (h)	Conversion (%)	RSCN (%)	R-NCS (%)
Cl	SiO_2	3	2	98	70	30
Cl	SiO_2	3	20	98	28	72
Cl	SiO_2	10	3	97	78	22
Cl	SiO_2	10	20	98	37	43
Cl	AlO_3	2	2	98	79	21
Cl	AlO_3	2	20	98	53	47
Cl	AlO_3	10	3	93	75	25
Cl	AlO_3	10	20	97	70	30
Cl	CaF_2	0.5	3	98	75	25
Cl	CaF_2	0.5	15	97	71	29

Figure 4.4 Effect of loading on selectivity of thiocyanate formation (see Section 4.2.1).

of the thiocyanate to the isothiocyanate occurred with this reagent, and also with the 2 mol/g alumina-supported reagents. The 0.5-mmol/g CaF_2 reagent on the other hand, gave rapid thiocyanation, but no isomerization with time. Again for supported reagents where the isomerization reaction was a problem, increasing the loading to 10 mmol/g helped to suppress the side reaction, at the expense of a drop in reaction rate (Fig. 4.4). Because of the course of these reactions it was found that the thiocyanate product could be separated from the isothiocyanate by-product by washing with hexane the isothiocyanate product off the supported reagent. The thiocyanate could then be obtained by ether washing. This observation was extended to provide a novel system to carry out the reaction and separate the products.

Similar problems of isomerization exist in the preparation of tertiary thiocyanates. Reactions to produce such materials were found to be quite successful using the supported reagents described. It was also found that the isothiocyanate formed was almost completely desorbed into solution, whereas the thiocyanate was present mostly on the solid surface. Thus, a column approach was used, where the reaction and the separation take place in one step. The combination of 2 mmol/g KSCN–alumina and 3 mmol KSCN–SiO_2 was found to be the most efficient for reaction and separation (Fig. 4.5):

Another application of KSCN-supported reagents is the two-phase addition to double bonds. Ando et al.[29] applied a variety of supported thiocyanates to the following scheme:

$$R{-}CH{=}CH{-}R' \xrightarrow{I_2/M^+X^-} R{-}\underset{\displaystyle I}{\underset{|}{CH}}{-}\underset{\displaystyle X}{\underset{|}{CH}}{-}R'$$

X = SCN, N_3, OAc

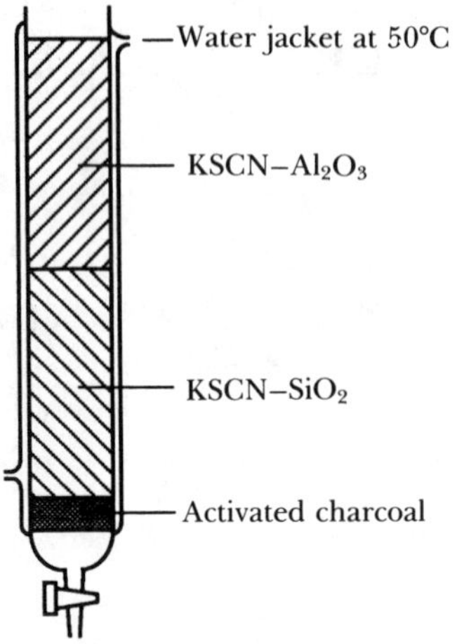

Elution of *t*-butyl bromides with hexane over 2–3 h at 50°C gave an isolated yield of *t*-butyl thiocyanate of 53% in a purity of 97%.

Figure 4.5 Preparation of *t*-butyl thiocyanate using a column reactor.

Such reactions have been carried out using phase transfer catalysis.[30] They involve reaction of the double bond with iodine in the organic solution, followed by nucleophilic attack on the resultant electrophile at the interface with the aqueous solution of necleophile.

Thus, Clark et al.[31] obtained a mixture of isomers with the following system (Fig. 4.6). In contrast with this, all the supported thiocyanates gave very high yields with the exclusive production of the iodothiocyanate (Fig.

Figure 4.6 Two-phase addition to double bonds using a PTC system.

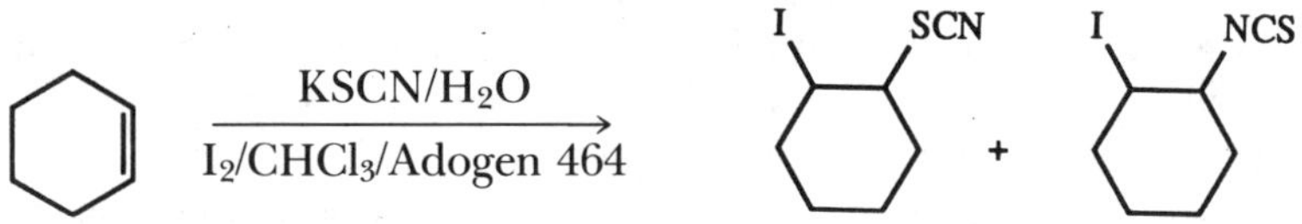

Figure 4.7 Two-phase addition to double bonds using supported reagents.

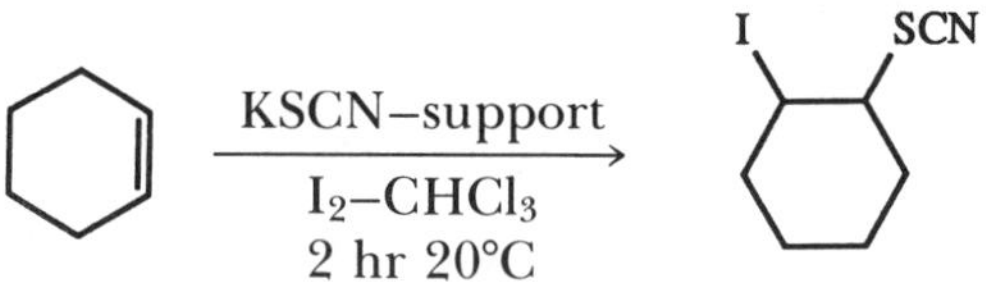

Support	Loading	Yield
CaF_2	0.5	96
Al_2O_3	2	82
SiO_2	3	86

4.7). Several other substrates gave very good conversions to the thiocyanate product.

Other nucleophiles were also successful. Azide ion gave the iodo azide in high yield, and was most successful on alumina rather than on CaF_2. This is probably because azide, being less reactive than thiocyanate in these reactions, requires better disperson/activation. This is available from the hydroxylated alumina surface (to a greater extent) than from the non hydroxylated CaF_2, making the NaN_3–Al_2O_3 superior to the NaN_3–CaF_2 reagent. For the very active SCN^- ion, all supports used gave very high yields and the ease of product recovery from the CaF_2 (which adsorbs the product less strongly than Al_2O_3 or SiO_2) means that it is recovered in higher yield.

4.2.5. CuSCN–Charcoal

Another supported thiocyanate reagent has been reported.[31] This is copper(I) thiocyanate supported on charcoal. Copper(I) reagents supported on charcoal have been shown to be useful in aromatic substitution reactions. Thus, CuI–charcoal converts bromobenzene to iodobenzene.[32] Similarly CuCN–charcoal can be used to prepare benzonitrile from bromobenzene.[33] In a similar manner CuSCN–charcoal has been developed and has similar reactivity to these reagents.

Preparation. Because of the low solubility of Cu(I) salts in most solvents, precipitation has been used to prepare the Cu-supported reagents. CuX will dissolve in solutions of KX in the following equilibrium:

$$KX + CuX \rightleftharpoons K^+CuX_2^- \qquad 4.5$$

At high concentration the equilibrium lies to the right and CuX is solubilized by adduct formation. Addition of water pushes the equilibrium over to the left and CuX is precipitated. This is one method for the purification of Cu(I) species.

If a support is added to a solution of $KCu(SCN)_2$, and water is added *slowly* with vigorous stirring, a supported CuSCN reagent is obtained. Thus, CuSCN–charcoal was prepared in this manner, filtered, washed, and dried to give a 3.3 mmol/g loading.

This reagent proved to be active in aromatic substitutions. Interestingly, the same reagent preparation carried out rapidly produced an inactive reagent.

Reactions. CuSCN–charcoal reacts with aryl bromides and iodides to give good yields of arylthiocyanates in non polar solvents such as *t*-butylbenzene (Fig. 4.8):

Substrate	Temperature (°C)	Time (h)	Conversion (%)
PhI	150	14	72
PhBr	150	25	43
2-$CH_3C_6H_4I$	150	25	69
4-$CH_3C_6H_4I$	150	27	72
2-$NO_2C_6H_4I$	150	7	84
4-$NO_2C_6H_4I$	150	24	0

Figure 4.8 Formation of aromatic thiocyanates using CuSCN–C.

Therefore, high yields of products are obtained in all cases but two.

Bromobenzene is less reactive than iodobenzene. This allows side reactions (formation of sulfide and disulfide products) to occur, thus lowering the yield of thiocyanate.

Reactions of iodotoluenes give similar yields to iodobenzene, but have slightly slower rates of reaction. Ortho and para isomers both react at the same rate.

2-Nitroiodobenzene reacts considerably faster than iodobenzene, and a very good conversion is obtained in a short time. 4-Nitroiodobenzene, on the other hand, gives no reaction whatsoever. This is presumably because the nitro group interacts strongly with the supported reagent surface and holds the iodine center away from the surface, stopping the reaction.

Analysis. FTIR spectra of these reagents have been recorded. As was the case for the anion in the other Cu(I) reagents, the SCN bond is shifted to a higher wave number in the supported reagent. Thus, unsupported CuSCN has a band at 2156 cm^{-1}. The active supported reagent (by *slow* precipitation) has this band, and also a shoulder at 2173 cm^{-1}. The inactive reagent, prepared by fast precipitation, does not display this second band. This second band disappears on reaction, and only the unshifted band is present in an used reagent. The active species is likely to correspond to the band at 2173 cm^{-1}, which probably corresponds to a bridged structure such as in Figure 4.9.[34] Thus, CuSCN–charcoal is a highly active material because of the formation of bridged species that are well dispersed on the charcoal surface.

Figure 4.9 Possible structure of the active species in CuSCN–C.

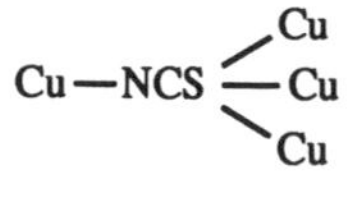

4.3. Friedel–Crafts Alkylation and Acylation Using Aluminosilicate-based Catalysts

4.3.1. Disadvantages of Conventional Friedel–Crafts Catalysts

Historically, the most important catalyst of Friedel–Crafts reactions is the Lewis acid aluminium trichloride. The use of this material, however, is fraught with serious problems: $AlCl_3$ is a corrosive irritant that decomposes in the presence of water; it is often required in greater than stoichometric quantities and frequently results in unselective reactions; furthermore, it may not be reused. Most importantly aluminum residues often remain in reaction products and large volumes of HCl gas and acidic effluent are created, both increasingly unacceptable in the industrial scale.

It is with these factors in mind that a number of researchers have attempted to design heterogeneous replacements for $AlCl_3$ in Friedel–Crafts chemistry. This case study focuses on some examples of aluminosilicate-based supported reagents.

4.3.2. Alkylation

The catalytic alkylation of benzene in the vapor phase has been carried out using an Egyptian natural aluminosilicate clay activated with $ZnCl_2$.[35] The catalyst was prepared by heating the clay to 375°C for 4 h before soaking it overnight in a 10% solution of $ZnCl_2$. The material was then dried and further activated in air at 500°C for 4 h. Benzene and isobutyl chloride were then flowed through the reactor over the catalyst at temperatures between 200 and 350°C. The $ZnCl_2$-treated clay was found to be catalytically active (the untreated clay was inactive) giving a maximum yield at 250°C. At this temperature the selectivity for mono-butylation was 87%. Interestingly, on increasing the reaction temperature from 250 to 300°C there is a switch from favoring the tert-butylbenzene product (over sec) to the reversal of this (tert : sec = 67 : 32 at 250°C and 27 : 73 at 300°C).

More recently Laszlo and Mathy[36] have reported the catalysis of Friedel–Crafts alkylations by a montmorillonite clay doped with transition metal cations. In this (and in all further examples) the reactions are carried out in the condensed phase. The transition metal ion-exchanged clays may be used for Friedel–Crafts alkylations with halides, alcohols, and alkenes. The catalysts exhibit favorable qualities over $AlCl_3$ catalysis with smaller quantities of catalyst being required along with a decrease in the amount of polyalkylation and improved regioselectivities. The choice of transition metal exchanged is found to be important and different ions favor different substrates. For example, the reaction of benzene with benzyl chloride to give diphenylmethane is best accomplished using Fe(III), Zn(II), Cu(II), or Zr(IV) doped montmorillonite K10. Laszlo and Mathy comment that the K10–Ti(IV) system is the catalyst of choice resulting in a "best compromise between reactivity and selectivity." The K10–Cu(II) catalyst performs poorly in alkylations with alkenes and K10–Co(II) and K10–Zn(II) are not suitable for alcohol alkylations.

It is important to note that the established trend of Lewis activity for homogeneous alkylations is not reflected in the reactions described previously.[37] Also the catalyst may be reused several times.

These catalysts were prepared by stirring the clay in a 1M solution of the appropriate transition metal chloride for 24 h, centrifuging and washing the clay catalyst until no further Cl^- can be removed. The catalyst is then activated overnight in air at 120°C and finely ground.

More highly active catalysts than those described have been prepared by supporting a transition metal salt onto the clay (by evaporation of the solvent) instead of ion exchanging.[38] For example, using a 1 mol/g loading of $ZnCl_2$ on K10 (Clayzic), activated at 280°C, stirred in 0.1 mole $PhCH_2Cl$ with 2 mole PhH at room temperature, 100 % conversion (80% isolated diphenylmethane) is achieved in less than 0.25 h. Several other alkylation reactions have been reported using such catalysts.[39]

Methylphenylmethanes have also been synthesized by the condensation of toluene in the presence of bromine and a bentonite earth catalyst.[40] This represents a two-step reaction: first, a free radical bromination of toluene followed by the Friedel–Crafts reaction of benzyl bromide on toluene.

The effect of acid treatment on the activity of clay catalyst supports for Friedel–Crafts reactions has been studied.[41] The $ZnCl_2$-supported montmorillonite described earlier[38] was studied under various acid treatments of a clay by surface area analysis, cation exchange capacity determination, powder X-ray diffraction, and thermal analyses.

The optimum state of the clay was determined to be between the laminar montmorillonite structure and the amorphous silica. The optimum state was defined by X-ray diffraction characterization correlated with time of exposure of the clay to mineral acid. Brown has further concluded that there is a synergystic interaction between the $ZnCl_2$ and the clay.

4.3.3. Acylation

The catalytic acylation of toluene by carboxylic acids using zeolites has been described.[42] In this case a high degree of regioselectivity was observed and explained by the shape selectivity imposed by the well-defined shape and size of zeolite channels in which the reaction takes place. Such a reaction was found to proceed by the formation of an acylium ion RCO^+ by the determination of the effect of structure on reactivity.[43]

A comparison of such reactions catalyzed by ion-exchanged montmorillonites and zeolites showed a dependence on the cation and catalyst structure.[44] Al^{3+} and H^+ were the most efficient cations from the montmorillonite catalysis with short and medium chain length acids, more meta-isomer was formed over the clay than over the zeolite. Exclusive para-acylation was achieved with long chain length acids (C-16–C-20) over zeolites. The two-dimensional clay structure imposed less steric constraints on the reaction than the three-dimensional zeolitic one.

Gauthier et al.[45] concluded that "The acidity and the catalytic activity of cation exchanged zeolites are strongly influenced by the nature of the exchanged cation and by the extent of the exchange, the most efficient catalysts being the rare earth-exchanged zeolites".

Using clay catalysts, Cornelis et al.[46] have reported an Fe(III)-exchanged montmorillonite as giving excellent yields in acylation reactions using benzoyl chloride or benzoic anhydride. Here, near-quantiative results were achieved with anisole and mesitylene in 15 min at 140–160°C.

A number of catalysts have been described in this Friedel–Crafts case study. Both clays and zeolites have been used to form catalytic materials overcoming most of the disadvantages of aluminium trichloride and thus providing environmentally friendly alternatives to that homogeneous catalyst.

REFERENCES

1. J. H. Clark, Chem. Rev., (1980), **80,** 429.
2. T. Ando and J. Yamawaki, Chem. Lett., (1979), 45.
3. J. Yamawaki and T. Ando, Chem. Lett., (1979), 755 and (1980), 522.
4. J. Yamawaki, T. Ando and T. Hanafusa, Chem. Lett., (1981), 1143.
5. T. Ando, J. Yamawaki, T. Kawate, S. Sumi and T. Hanafusa, Bull. Chem. Soc. Jpn., (1982), **55,** 2504.
6. J. H. Clark, D. G. Cork and M. S. Robertson, Chem. Lett., (1983), 1145.
7. P. Laszlo and P. Pennetreau, Tetrahedron Lett., (1985), 2645.
8. J. H. Clark, D. G. Cork and H. W. Gibbs, J. Chem. Soc. Perkin 1, (1983), 2253.
9. S. Kaluz and S. Toma, Collec. Czech. Chem. Commun., (1986), **51,** 2199.
10. J. M. Méloz, F Texier-Boullet and A. Foucaud, Tetrahedron Lett., (1988), **44,** 2215.
11. D. Villemin, J. Chem. Soc., Chem. Commun., (1983), 1092.
12. F. Texier-Boullet, D. Villemin, M. Ricard, H. Moison and A Foucaud, Tetrahedron, (1985), **41,** 1259.
13. F. Texier-Boullet and M. Lequitte, Tetrahedron Lett. (1986), 3515.
14. J. M. Méloz, F. Texier-Boullet and A. Foucaud, Tetrahedron Lett., (1986), 493.

15. S. K. Suri, K. Thangaraj, A. K. S. Bhujanga Rao, R. S. Prasad, C. Gurdu Rao and S. P. Bhatnagar, Chem, and Ind., (1988), 94.
16. D. Villemin, Chem, and Ind., (1985), 166.
17. J. H. Clark, A. J. Hyde and D. K. Smith, J. Chem. Soc. Chem. Commun., (1986), 791.
18. J. Ichihara, T. Matsuo, T. Hanafusa and T. Ando, J. Chem. Soc. Commun., (1986) 793.
19. H. Liu, P. Wang and P. Sun, J. Fluorine Chem., (1989), 429.
20. T. Ando, J. H. Clark, K. G. Cork and T. Kimura, Bull. Chem. Soc. Jpn., (1986). **59,** 3281.
21. J. H. Clark, E. M. Goodman, D. K. Smith, S. J. Brown and J. M. Miller, J. Chem. Soc., Chem. Commun., (1986), 657.
22. T. Ando, S. J. Brown, J. H. Clark, D. G. Cork, T. Hanafusa, J. Ichihara, J. M. Miller and M. S. Robertson, J. Chem. Soc. Perkin Trans. 2, (1986), 1133.
23. L. M. Weinstock, J. M. Stevenson, S. A. Tomellini, S. H. Pan, T. Utne, R. B. Jobron and D. F. Reinhold, Tetrahedron Lett., (1986), 3845.
24. T. Ando, J. H. Clark, D. G. Cork, T. Hanafusa, J. Ichihara and T. Kimura, Tetrahedron Lett., (1987), 1421.
25. B. Schülze, M. Muhlstadt and I. Schubert, Z. Chem., (1979), **19,** 41.
26. M. O. Brimeyer, A. Methrotra, S. Quici, A. Nigam and S. L. Regan, J. Org. Chem., (1980), **45,** 4254.
27. M. Kodomari, T. Kuzuoka and S. Yoshitomi, Synthesis, (1983), 141.
28. T. Ando, J. H. Clark, D. G. Cork, M. Fujita and T. Kimura, J. Org. Chem., (1987), **52,** 681.
29. T. Ando, J. H. Clark, D. G. Cork, M. Fujita and T. Kimura, J. Chem. Soc., Chem. Commun., (1987), 1301.
30. P. D. Woodgate, H. H. Lee, P. S. Rutledge and R. C. Cambie, Synthesis, (1977), 462.
31. J. H. Clark, C. W. Jones, C. V. A. Duke and J. M. Miller, J. Chem. Soc., Chem. Commun., (1989), 81.
32. J. H. Clark and C. W. Jones. J. Chem. Soc., Chem. Commun., (1987), 1409.
33. J. H. Clark, C. V. A. Duke, S. J. Brown and J. A. Miller, J. Chem. Soc., Chem. Commun., (1986), 877.
34. M. Kabesova, M. Dunay-Jurco, M. Serator and J. Garay, Inorg. Chim. Acta., (1976), **17,** 161.
35. B. E. M. Hassan, E. A. Sultan, F. M. Tawfik and S. M. Sappah, Egypt, J. Chem. (1985), **28,** 93.
36. P. Laszlo and A. Mathy, Helv. Chim. Acta. (1987), **70,** 577.
37. A. P. Kybett, D. Phil. Thesis, York, (1988).
38. J. H. Clark, A. P. Kybett, D. J. Macquarrie, S. J. Barlow and P. Landon, J. Chem. Soc., Chem. Commun., (1989), 1353.
39. S. J. Barlow, J. H. Clark, M. R. Darby, A. P. Kybett, P. Landon and K. Martin, J. Chem. Res. (S), (1991), 74.
40. M. Salmon, E. Angeles and R. Miranda, J. Chem. Soc., Chem. Commun., (1990), 1188.

41. C. N. Rhodes, M. Franks, G. M. B. Parkes and D. R. Brown, J. Chem. Soc., Chem. Commun., (1991), 804.
42. B. Chiche, A. Finiels, C. Gauthier and P. Geneste, J. Org. Chem., (1986), **51,** 2128.
43. B. Chiche, A. Finiels, C. Gauthier and P. Geneste, App. Catal., (1987), **30,** 365.
44. B. Chiche, A. Finiels, C. Gauthier, P. Geneste, J. Graille and D. Pioch, J. Mol. Catal., Chem., (1987), **42,** 229.
45. G. Gauthier, B. Chiche, A. Finiels and P. Geneste, J. Mol. Catal., (1989), **50,** 219.
46. A. Cornelis, A. Gerstmans, P. Laszlo, A. Mathy and I. Zieba, Catal. Lett., (1990), **6,** 103.

Appendix 1
Organic Synthesis Using Supported Reagents

Oxidation and Reduction

Reaction Example	Supported Reagent	Comments	Reference
$C_6H_5CHO \rightarrow$ $\rightarrow C_6H_5CH_2OH + C_6H_5CO_2H$ (Cannizzaro)	$Na_2S_2O_3$–alumina	Na_2S–alumina or Na_2SO_3–alumina also effective. Acid adsorbed on support for useful product separation.	6
Cyclododecylethanoate → 5-ethoxycarbonyl dodecanone	Substrate–silica gel	Regioselective oxidation achieved by treating substrate-support with ozone	10
$PhCH{=}CHNO_2 \rightarrow PhCH_2CH_2NO_2$	$NaBH_4$ in presence of silica gel	Near quantitative silica gel-assisted reduction of nitrostyrenes	11
$CH_3(CH_2)_5CH(NO_2)CH_2CH_2COCH_3 \rightarrow$ $\rightarrow CH_3(CH_2)_5COCH_2CH_2COCH_3$	$KMnO_4$–silica gel	Nitroketone oxidation to diketone : Nef transformation	17
 $\rightarrow HOOC(CH_2)_4COOH$	$KMnO_4$–silica gel	C=C bond cleavage	20
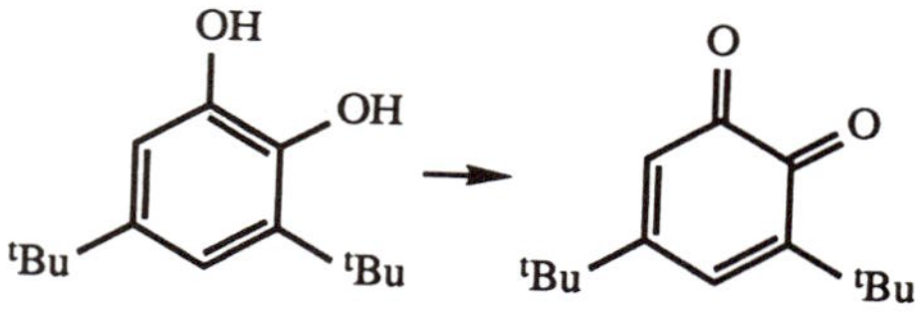	$FeCl_3$–silica	Comparison of various support materials and catalyst preparations	21
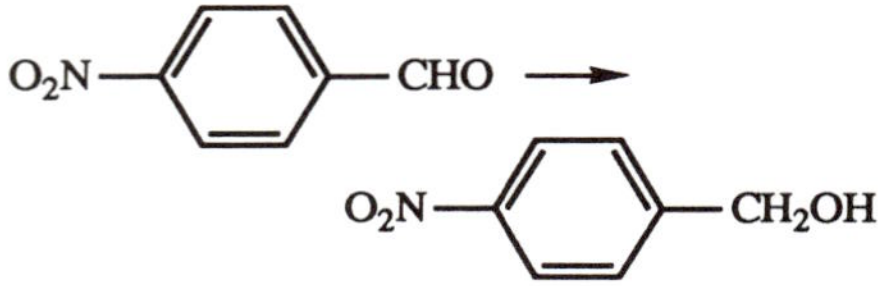	Na_2S–alumina	Selective reduction of aldehydes to alcohols	23

$CH_2{=}CHCH(OH)(CH_2)_5CH_3 \rightarrow CH_2{=}CHCO(CH_2)_5CH_3$	$KMnO_4$–bentonite	Selective oxidation of unsaturated alcohols	31
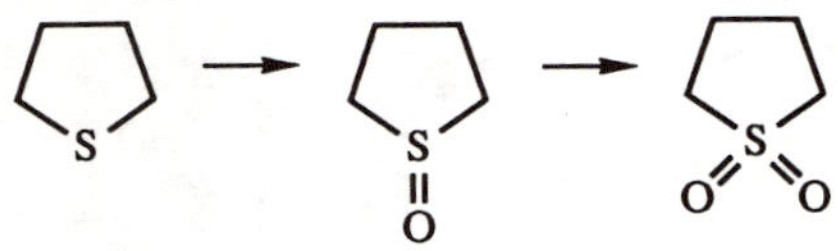	$NaIO_4$–aluminia		32
$PhCH_2OH \rightarrow PhCHO$	$Zn(NO_3)_2$ or $Cu(NO_3)_2$–silica gel	Also oxidation of secondary alcohols	34
$Ph_2CHOH \rightarrow PhCO$	$KMnO_4$–molecular sieves	Various molecular sieves and montmorillonites tested as permanganate supports	39
$(C_8H_{17}OH \rightarrow C_7H_{15}CHO$	CrO_2Cl_2–silica–alumina	Selective oxidation of primary and secondary alcohols	41
$PhCHOHCH_2CH_2OH \rightarrow PhCOCH_2CH_2OH$	$K_2Cr_2O_7$–alumina	Preferential oxidation of benzylic alcohols	42
$PhCOCH_3 \rightarrow PhCH_2CO_2CH_3$	$T1(NO_3)_3$–K10 montmorillonite	Oxidative rearrangements via oxythallation	46
$EtCHOHCH_2NO_2 \rightarrow EtCOCH_2NO_2$	CrO_3–montmorillonite	Production of nitroketones	47
$(CH_3)_2CHOH \rightarrow (CH_3)_2CO$	$Cu(NO_3)_2$–silica	Copper or zinc nitrates for oxidation of secondary alcohols and primary benzyl alcohols	54
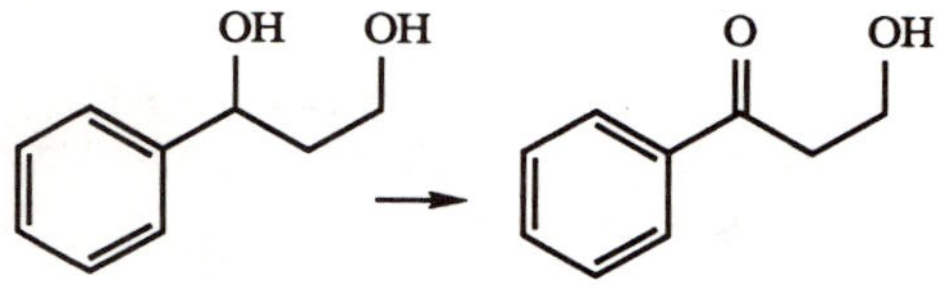	$BaMnO_4$–alumina–$CuSO_4$	Selective alcohol oxidation by manganate	62
$C_9H_{19}CH_2OH \rightarrow C_9H_{19}CHO$	H_2CrO_4–silica	Supported chromic acid for oxidation of 1° and 2° alcohols	63

Reaction Example	Supported Reagent	Comments	Reference
$CH_2OH(CH_2)_2CH_2OH \rightarrow$ 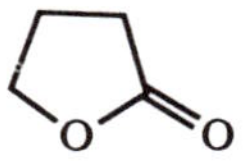	Pd–zeolite K-L	This supported Pt°–zeolite is highly selective for the oxidative esterification	64
$PhCHO \rightarrow PhCO_2H$	$KMnO_4$–silica	Also tested $AlPO_4$, BPO_4, and sepiolite as supports	65
$PhCH_2OCH_3 \rightarrow PhCHO$	$Cu(NO_3)_2$–silica	Convenient oxidative cleavage of ethers	66
$Ph_3CSH \rightarrow Ph_3CH + (Ph_3C)_2$	$Na(C_2H_5)_3BH$–alumina–$FeCl_2$	Desulfurization using co-supported borohydride and iron II	73
$CH_3(CH_2)_3CH{=}CH_2 \rightarrow CH_3(CH_2)_4CH_3$	$Rh(PPh_3)_2^+$–clay	Hydrogenation using mica-type silicates	77
$PhNO_2 \rightarrow PhNH_2$	$[Co(bipy)_3]^{3+}$–clay	bipy = 2,2′-bipyridyl. This hectorite material catalyzes the BH_4^- reduction	78
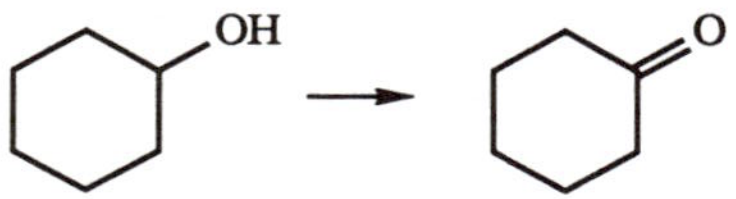	$Fe(NO_3)_3$–clay	Air oxidation via nitrous ester	82
$PhCH{=}CH_2 \rightarrow PhCH_2CH_3$	$PdCl_2$–clay	Consistent reduction rate over 1000 tunovers	84
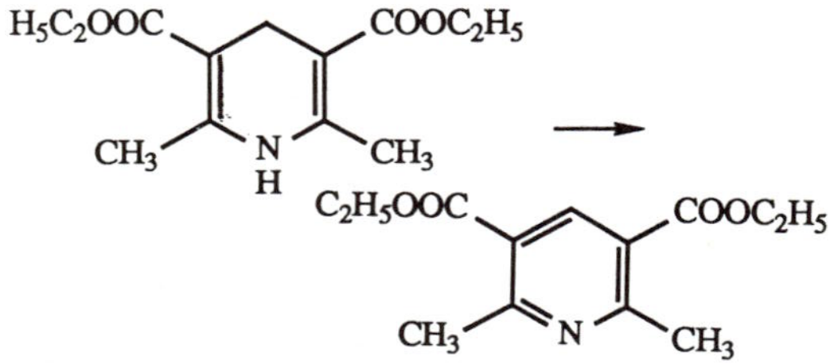	$Cu(NO_3)_2$–clay	Aromatization using "clayfen" or "claycop"	91

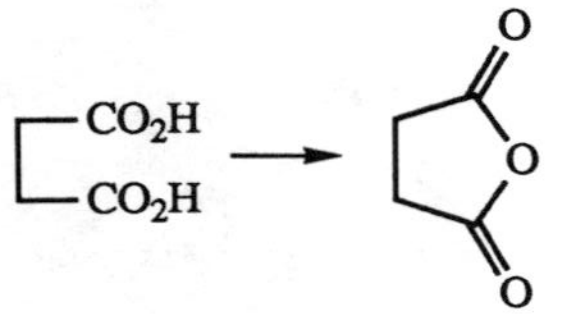	Al^{3+}–clay	Clay and zeolite catalyzed cyclic anhydride formation	93
$C_5H_{11}OH \rightarrow C_3H_7CH{=}CH_2 + C_5H_{11}OC_5H_{11}$	Al^{3+}–clay	Dehydration comparing exchanged and pillared clays	94
$C_6H_{13}CH_2OH \rightarrow C_6H_{13}CHO$	$Fe(NO_3)_3$–clay	"clayfen" for alcohol oxidation	95
$C_6H_5SH \rightarrow C_6H_5SSC_6H_5$	$Fe(NO_3)_3$–clay	"clayfen" for the coupling of thiols	96
$2CH_4 \rightarrow C_2H_6$	$Pb(NO_3)_2$–silica–$Mn(NO_3)_2$	Synthetic kentrolite for selective oxidation of methane	97
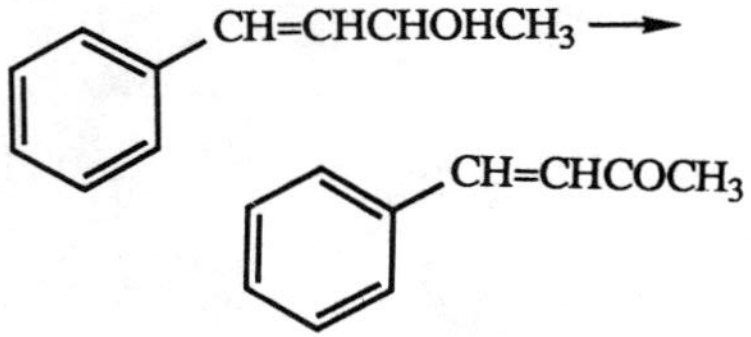	$KMnO_4$–clay	Selective oxidation of 2° alcohols in the presence of C=C	100
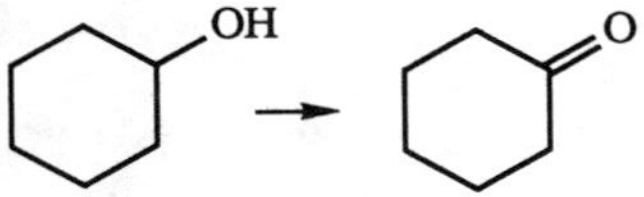	CrO_3–silica–C_5H_5N	Pyridinium chromate–silica for 1° and 2° alcohol oxidation	102
$Ph_2CHOH \rightarrow Ph_2CO$	$KMnO_4$–zeolite	Permanganate-coated molecular sieve pellets	103
$Ph_2CHOH \rightarrow Ph_2CO$	Ag_2CO_3–celite	Mechanistic studies	104
$Ph_2CH_2 \rightarrow Ph_2CO$	$K_2Cr_2O_7$–alumina	Truly catalytic aerial oxidations	106

Reaction Example	Supported Reagent	Comments	Reference
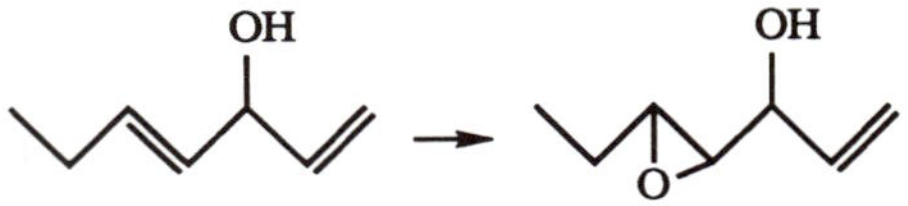	Vanadium-pillared montmorillonite	Regioselective epoxidation	109
	Titanium-pillared montmorillonite	Catalytic asymmetric epoxidation	114
PhCHOH → Ph_2CO	$KMnO_4$–alumina	Oxidation of alcohols	107
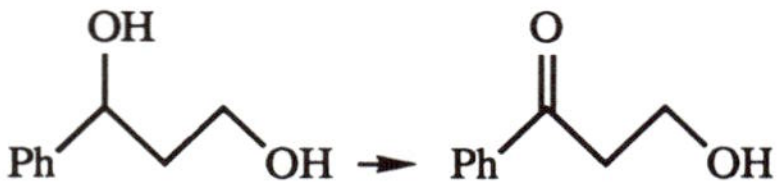	Chromia-pillared clay	Selective alcohol oxidation	111
$CH_3(CH_2)_4CH_2OH \rightarrow CH_3(CH_2)_4CHO$	$KCrClO_3$–alumina	Benzylic and 2° alcohols give highest yields	112

Addition to C-Heteroatom Multiple Bonds

$CHCl_3 + mO_2NC_6H_4CHO \rightarrow mO_2NC_6H_4CH(OH)CCl_3$	KF–alumina		7
$CH_3CH_2C(=N\text{—}NMe_2)(CH_2)_2CH_3 \rightarrow CH_3CH_2CO(CH_2)_2CH_3$	$Fe(NO_3)_3$–clay	"clayfen" for *N,N*-dimethylhydrazone cleavage	105

Addition to C—C Multiple Bonds

$MeOH + CH_2{=}C(CH_3)_2 \rightarrow MeOC(CH_3)_3$	Amberlyst 1S ion exchange resin or ion exchanged montmorillonite	Comparison of IER and montmorillonites for MTBE production	1
$CH_3(CH_2)_3CH{=}CH_2 + DBH \rightarrow CH_3(CH_2)_3CHFCH_2Br$	$AgF–CaF_2$	DBH:1,3-dibromo-5,5-dimethyl hydantoin (bromine source). Also with other Br or I sources for halofluorination	2
$PhCH{=}CH_2 + I_2 \rightarrow$ $\rightarrow PhCH(SCN)CH_2I$ $\rightarrow PhCH(N_3)CH_2I$ $\rightarrow PhCH(OAc)CH_2I$	 KSCN–alumina NaN_3–alumina KOAc–alumina	Selective two-phase additions to C═C. Comparison of various supports and substrates	5
$CH_3CH_2NO_2 + CH_3COCH{=}CH_2 \rightarrow CH_3COCH_2CH_2CH(NO_2)CH_3$	KF–alumina	Fluoride catalyzed Michael reactions	15, 17
$CH_3COCH_2COCH_3 + CH_2{=}CHCOCH_3$	^tBuOK–xonotlite or KF–alumina	Base-catalyzed Michael reactions	26
$Et_2NH + EtO_2CCH{=}CH_2 \rightarrow Et_2N(CH_2)_2CO_2Et$	Fe^{III}–montmorillonite	Acrylate acceptors in Michael reactions	56
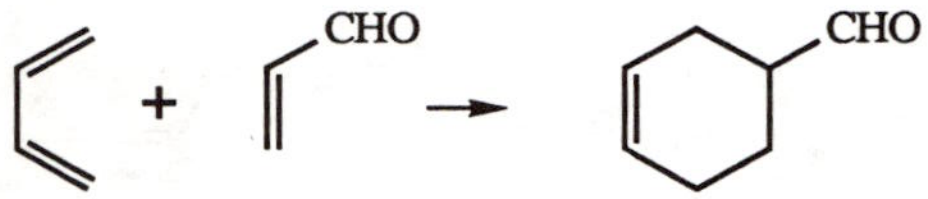	Fe^{III}–montmorillonite	Diels–Alder catalysis	60, 87
$CH_3NO_2 + CH_2C{=}CHCO_2C_2H_5 \rightarrow$ $\rightarrow CH_2(NO_2)CH_2CH_2CO_2C_2H_5$	KF–alumina	Michael additions of nitroalkanes	74

Reaction Example	Supported Reagent	Comments	Reference
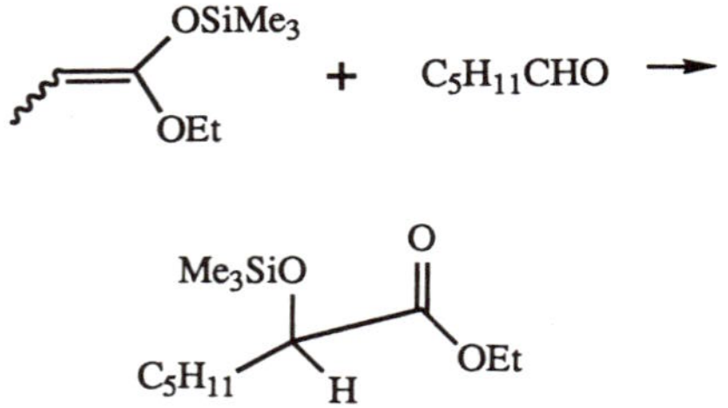	Al^{3+}–clay	Al^{3+}-exchanged montmorillonite for Aldol reactions	
$Me_2{=}CH_2 + MeOH \rightarrow (CH_3)_3COMe$	Al^{3+}–clay	Methanol addition with Al^{3+}-exchanged montmorillonite	81
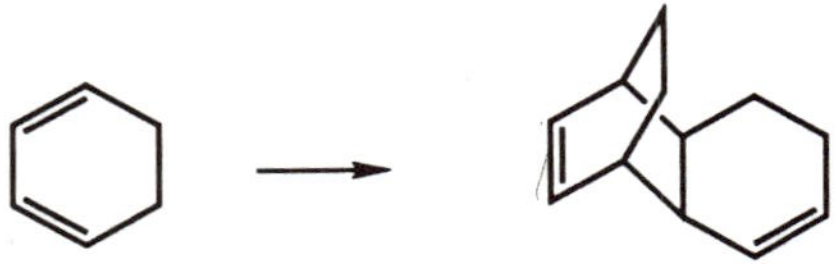	Fe^{3+}–clay–4-^{t}Butylphenol	4:1 endo:exo Diels–Alder	83
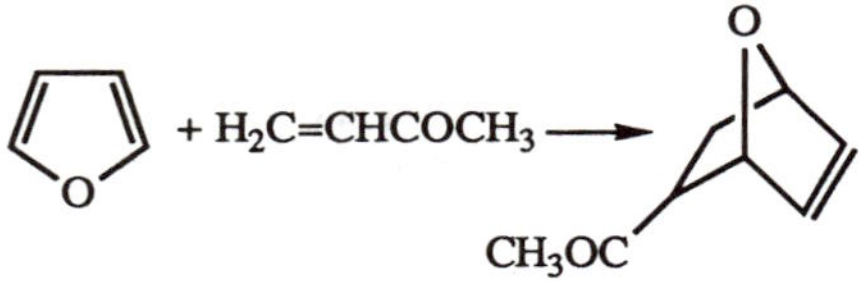	$AlCl_3$–clay	endo:exo ratio dependent on $AlCl_3$ or Fe^{3+}–clay	88
$CH_2{=}CH_2 + CH_3CO_2H \rightarrow$ $\rightarrow CH_3CO_2CH_2CH_3$	Al^{3+}–clay	Facile ester synthesis with Al^{3+}-exchanged montmorillonite	89
$PhCl + MeOH \rightarrow ClC_6H_4OMe$	H–mordenite	Vapor phase alkylation after treatment of support with $CFCl_3$	90

Intramolecular Rearrangement

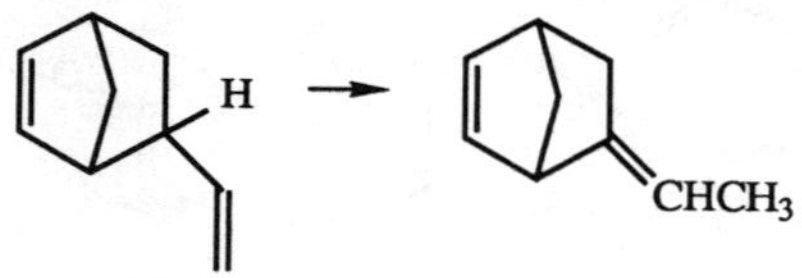	NaOH–alumina–Na	Solid superbase for catalytic isomerization	75
$nC_5H_{12} \rightarrow iC_5H_{12} + iC_4H_{10}$	$AlCl_3$–alumina	Solid superacid alkane isomerization	76
$ArCH_2CH{=}CH_2 \rightarrow ArCH{=}CHCH_3$	KF–alumina	alkene Isomerization	108

Aromatic Substitution

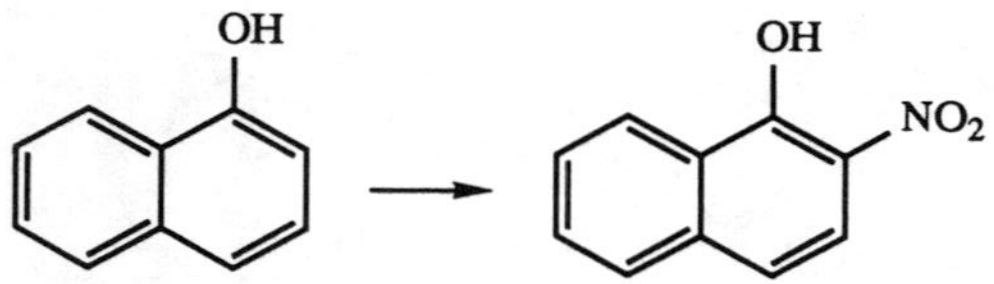	$(NH_4)_2Ce(NO_3)_6$–silica gel	Mononitration of arenes and hydroxynapthalenes	12
$PhI \rightarrow PhSCN$	CuSCN–charcoal	ArI and ArBr → ArSCN with no ArNCS	14
$PhBr \rightarrow PhI$	CuI–alumina or CuI–charcoal	ArBr and ArCl → ArI with no reverse reaction	16
$CH_3(CH_2)_{10}COOH + PhCH_3 \rightarrow$ $\rightarrow CH_3(CH_2)_{10}COC_6H_4CH_3$	CeNaY–zeolite	Friedel–Crafts acylation by zeolites	22

Reaction Example	Supported Reagent	Comments	Reference
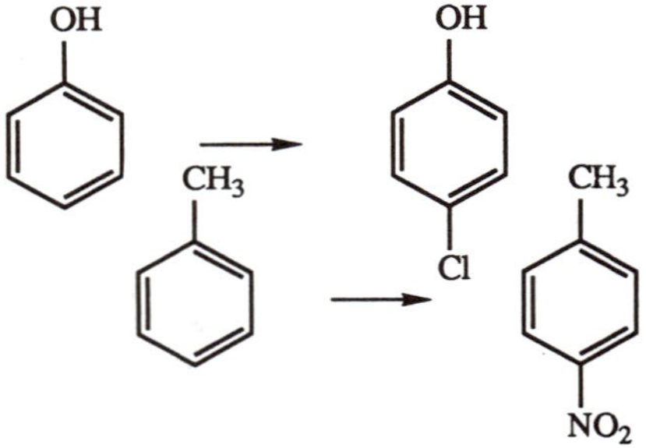	$CuCl_2$–alumina	Regioselective chlorination of phenols	24
	$Cu(NO_3)_2$–montmorillonite	Regioselective nitration of aromatic hydrocarbons	27
$C_6H_6 + C_6H_5CH_2Cl \rightarrow Ph_2CH_2$	Zr(IV)- or Ti(IV)-exchanged montmorillonite	Friedel–Crafts alkylations	28
$C_6H_6 + C_6H_5CH_2Cl \rightarrow Ph_2CH_2$	$ZnCl_2$–montmorillonite	Friedel–Crafts alkylations with increased rate, see Ref. 28	29
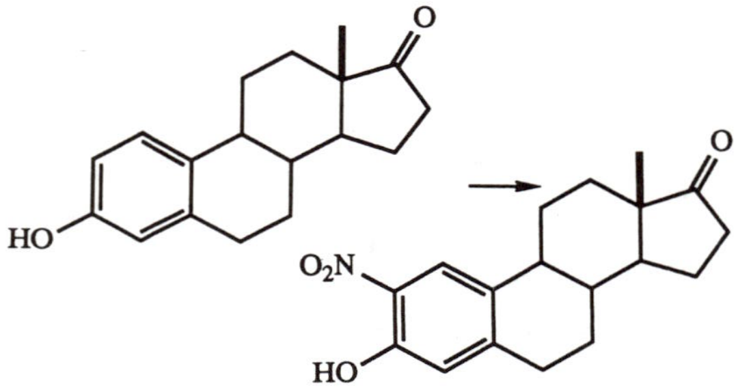	$Fe(NO_3)_3$–montmorillonite	Regioselective nitration of estrone	30

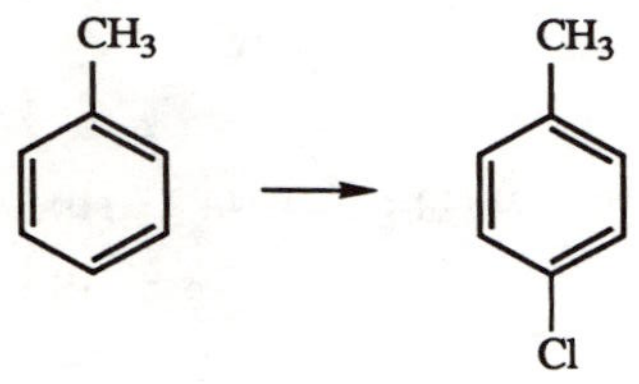	${}^{t}BuOCl$–zeolite	Paraselective monochlorination	40
$PhCH_3 \rightarrow Cl-C_6H_4CH_3$	${}^{t}BuOCl$–silica	A less regioselective reaction than that using ${}^{t}BuOCl$–zeolite (above)	43
$C_6H_6 \rightarrow C_6H_5NO_2$	$PhSO_3H$–silica	Vapor phase reaction with NO_2	45
$PhOH \rightarrow {}^{t}BuC_6H_4OH$	Na^+–zeolite	C-alkylation catalyzed by high acidity sites	49
$C_6H_6 + CH_2O \rightarrow Ph_2CH_2$	Na^+–zeolite	Formaldehyde condensation with benzene studied determining active acid sites	50
$C_6H_6 + PhCH_2Cl \rightarrow Ph_2CH_2$	$H_3PMo_{12}O_{40}$–silica	Silica-supported heteropolyacids for Friedel–Crafts alkylation and acylation	53
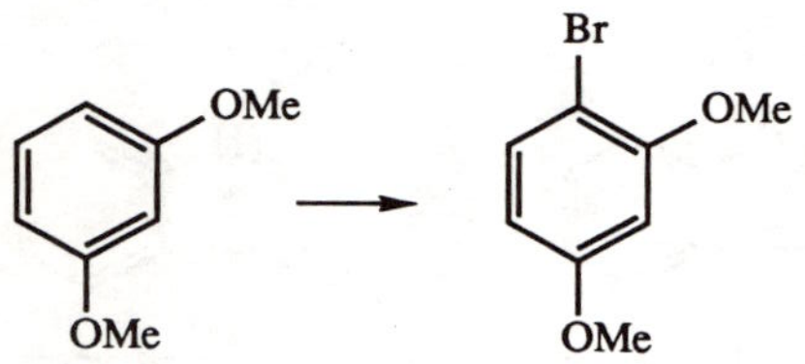	*N*-Bromosuccinimide–silica gel	Bromination of alkoxybenzenes	55
$PhOH \rightarrow Cl–C_6H_4OH$	$ClNR_2$–silica	High orthoselectivity in the chlorination of phenols	57

Reaction Example	Supported Reagent	Comments	Reference
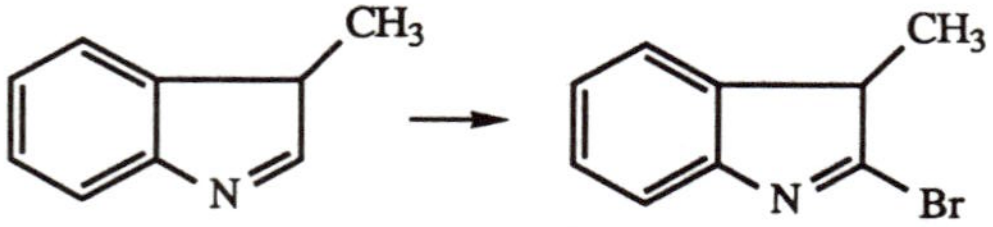	*N*-Bromosuccinimide–silica gel	Substantial regioselectivity in the bromination of indoles and benzimidazoles	58
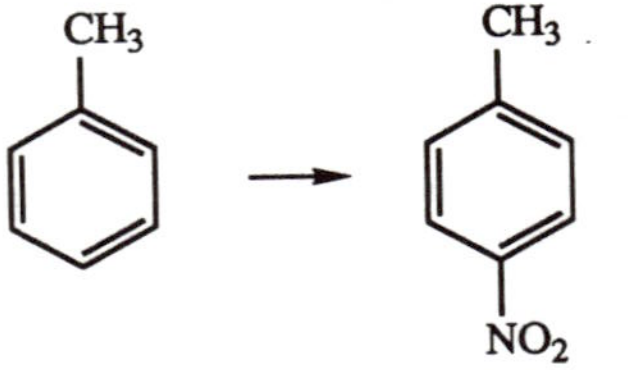	$PhCO_2NO_2$–zeolite	Paraselectivity achieved with large port mordenite zeolite	59
$PhCH_3 + (PhCO)_2O \rightarrow CH_3C_6H_4COPh$	H_2MoO_4–$Zr(OH)_4$	Calcination of this supported reagent gives a solid superacid catalyst	61
$PhCH_3 \rightarrow C_6H_4(CH_3)_2$	La^{3+}–pillared clay	La^{3+} modified pillared mica	79
$PhCH_3 + C_9H_{19}CO_2H \rightarrow CH_3C_6H_4COC_9H_{19}$	Al^{3+}–clay	Acylation with various ion-exchanged montmorillonites	85
$PhCH_3 + MeOH \rightarrow pCH_3C_6H_4CH_3$	Al–polyoxycation-pillared clay	High paraselectivity	86
$PhI \rightarrow PhSCN$	CuSCN–charcoal	No isothiocyanate contamination	99
$PhH + i\text{-}PrBr \rightarrow i\text{-}PrPh$	TiO_2–silica–alumina	Regenerable catalyst for alkylation with alkylhalides, alcohols, and alkenes	101

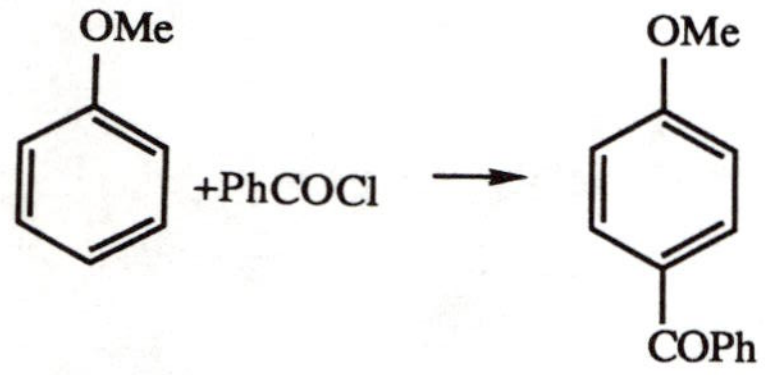	$ZnCl_2$–montmorillonite	Catalytic acylations with modified clays	98
$CH_3Ph \rightarrow CH_3C_6H_4Cl$	$FeCl_3$–clay	Chlorination in presence of metachloroperbenzoic acid. Also chlorination of anisole	110

Elimination

$EtNH_2 \rightarrow Et_2NH + NH_3$	Ion exchanged montmorillonites	Wide range of montmorillonites studied for the elimination of NH_3 from primary or cyclic amines	9
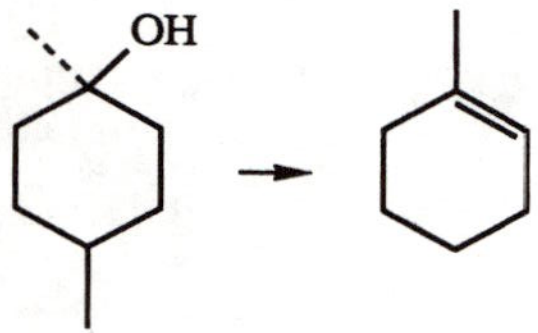	$FeCl_3$–silica gel	S.R. also converts epoxides to diols	33
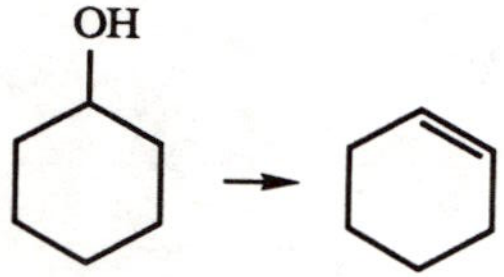	$Fe_2(SO_4)_3$–silica	Dehydration using silica-supported $Fe_2(SO_4)_3$, $CuSO_4$, and $NaHSO_4$	69

Reaction Example	Supported Reagent	Comments	Reference
Aliphatic Substitution			
$MeOH + HOC(CH_3)_3 \rightarrow$ $\rightarrow MeOC(CH_3)_3 + H_2O$	Amberlyst 1S ion-exchange resin or ion-exchanged montmorillonite	Comparison of I.E.R. and montmorillonites	1
$PhCH_2Br \rightarrow PhCH_2F$	$AgF–CaF_2$	Also applied to alkylchlorides or iodides	2
$PhCH_2Cl \rightarrow PhCH_2SCN$	KSCN–silica, alumina, or CaF_2	Comparison of various supports and SCN vs. NCS selectivity	3
$PhOH + MeOH \rightarrow PhOMe$	KF or CsF–alumina	O-alkylation of phenols and alcohols enhanced by supported F^-	4
$nC_8H_{17}Br \rightarrow nC_8H_{17}OAc$	Alumina	The essential role of water in this reaction is described	8
$PhCH_2Br \rightarrow PhCH_2F$	$KF–CaF_2$	Also applied to RCOCl	13
$PhCH_2Cl \rightarrow PhCH_2CN$	KCN–alumina	IR studies used to optimize supported reagent preparation	18
$CH_3(CH_2)_{14}CH_3 \rightarrow$ $CH_3CH_2CH_3 + C_4H_{10}$	$AlCl_3$–silica, alumina, zeolite, or B_2O_3	Catalytic cracking by solid acids	19
$nC_6H_{17}Br \rightarrow nC_6H_{17}SCN$	KSCN–silica gel	Also benzyl chlorides reacted	25
$nC_7H_{15}OH \rightarrow nC_7H_{15}OCH_3$	Me_2SO_4 and alumina	Also selective monomethylation of diols by adsorption of substrate on alumina	35

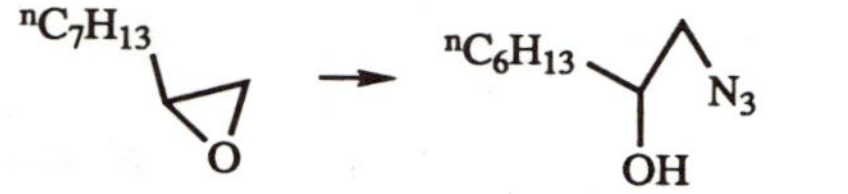	NaN_3–zeolite or alumina	Factors affecting reactivities discussed	36
$nC_{10}H_{21}OH + PhCH_2Cl \rightarrow nC_{10}H_{21}\text{–O–}CH_2Ph$	CsY zeolite	Various zeolites for the benzylations	37
$nC_4H_9Br \rightarrow nC_4H_9CN$	NaCN–alumina		38
$PhCOCl \rightarrow PhCOCN$	KCN–amberlite resins	Several inorganic supports were not effective	44
$EtCHO + CH_3NO_2 \rightarrow EtCHOHCH_2NO_2$	KF–alumina	Condensation of aldehyde and nitroalkane	47
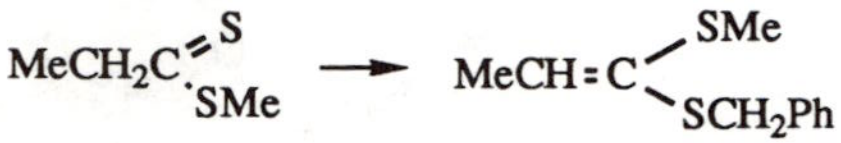	KF–alumina	Benzylchloride alkylation gives good *cis*-alkene selectivity	48
$PhOH \rightarrow PhO^tBu$	Na^+–zeolite	Varying Na^+ exchange influences C:O alkylation ratio	49
$HO(CH_2)_4OH \rightarrow HO(CH_2)_4O_2CCH_3$	$NaHSO_4$–silica	Selective monoacylation of 1,*n*-diols catalyzed by silica-supported metallic sulfates	51
$CHClF_2 \rightarrow CHF_3$	AlF_3–alumina	Fluorinated aluminas effect fluorine for chlorine exchange with $CHCl_xF_4$	52
$MeCOCl \rightarrow MeCOF$	KF–polystyrene	Nucleophilic activity of KF enhanced by support on cross-linked polymer	68
$PhCH_2Br \rightarrow PhCH_2CN$	KCN–alumina	Ultrasonic irradiation of a KCN and alumina mixture	70

Reaction Example	Supported Reagent	Comments	Reference
PhBr → PhI	CuI–alumina	Alumina- or charcoal-supported CuI for reverse halogenation	71
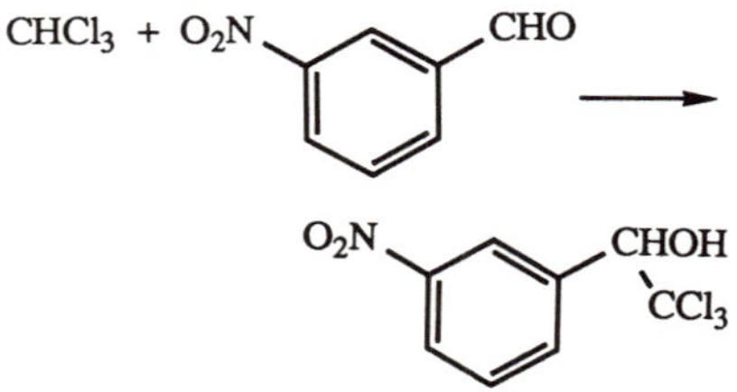	KOH–alumina	No Cannizaro side products	72
$CH_2Cl_2 + 2\ PhCH_2OH \rightarrow (PhCH_2O)CH_2$	NR_4^+–clay	Hydroxide promoted formation of symmetrical acetals with exchanged montmorillonites	92
$^tBuBr \rightarrow {}^tBuSCN + {}^tBuNCS$	$M(NCS)_2$-$CaCO_3$ or MF_2	M = Ba,Sr,Ca. Support-controlled switching of S:N nucleophilicity	67
$nC_8H_{17}Br \rightarrow CH_3COO$-$nC_8H_{17}$	CH_3COOK–alumina	Alkylation in "dry media" using microwave activation	113
PhCOCl → PhCOF	KF–alumina (SiN)	Surface-treated supports for nucleophilic substitutions. SiN: treatment with hexamethyldisilazane	115

REFERENCES TO APPENDIX 1

1. J. M. Adams, K. Martin, R. W. McCabe and S. Murray, Clays & Clay Miner., (1986), **34,** 597.
2. T. Ando, D. G. Cork, M. Fujita, T. Kimura and T. Tatsuno, Chem. Lett., (1988), 1877.
3. T. Ando, J. H. Clark, D. G. Cork, M. Fujita and T. Kimuro, J. Org. Chem., (1987), **52,** 681.
4. T. Ando, J. Yamawaki, T. Kawate, S. Sumi, T. Hanafusa, Bull. Chem. Soc. Jpn., (1987), **55,** 2504.
5. T. Ando, J. H. Clark, D. G. Cork, M. Fujita and T. Kimura, J. Chem. Soc., Chem. Commun., (1987), 1301.
6. T. Ando, J. H. Clark, D. G. Cork and T. Kimura, Chem. Lett., (1987), 1315.
7. T. Ando, J. H. Clark, D. G. Cork, T. Hanafusa, J. Ichihara and T. Kimura, Tetrahedron Lett., (1987), **28,** 1421.
8. T. Ando, T. Kawate, J. Yamawaki, and T. Hanafusa, Chem. Lett., (1982), 935.
9. J. A. Ballantine, J. H. Purnell, M. Rayanakorn, K. J. Williams and J. M. Thomas, J. Mol. Catal., (1985), **30,** 373.
10. A. L. J. Beckwith and T. Duong, J. Chem. Soc., Chem. Commun., (1978), 413.
11. A. K. Sinhababu and R. T. Borchardt, Tetrahedron Lett., (1983), **24,** 227.
12. H. M. Chawla and R. S. Mittal, Synthesis, (1985), 70.
13. J. H. Clark, A. J. Hyde and D. K. Smith, J. Chem. Soc., Chem. Commun., (1986), 791.
14. J. H. Clark, C. W. Jones, C. V. A. Duke and J. M. Miller, J. Chem. Soc., Chem. Commun., (1989), 81.
15. J. H. Clark, D. G. Cork and M. S. Robertson, Chem. Lett., (1983), 1145.
16. J. H. Clark and C. W. Jones, J. Chem. Soc., Chem. Commun., (1987), 1409.
17. J. H. Clark, D. G. Cork and H. W. Gibbs, J. Chem. Soc., Perkin Trans. 1, (1983), 2253.
18. J. H. Clark and C. V. A. Duke, J. Org. Chem., (1985), **50,** 1330.
19. R. S. Drago and E. E. Getty, J. Am. Chem. Soc., (1988), **110,** 3311.
20. J. T. B. Ferreira, W. O. Cruz, P. C. Viera and M. Yonashiro, J. Org. Chem., (1987), **52,** 3698.
21. T. Funabiki, T. Sugimoto and S. Yoshida, Chem. Lett., (1982), 1097.
22. B. Chiche, A. Finiels, C. Gauthier and P. Geneste, J. Org. Chem., (1986), **51,** 2128.
23. Y. Kamitori, M. Hojo, R. Masuda and M. Yamamoto, Chem. Lett., (1985), 253.
24. M. Kodomari, S. Takahashi and H. Watanabe, Nippon Kagaku Kaishi, (1987), 1753.
25. M. Kodomari, T. Kuzuoka, S. Yoshitomi, Synthesis, (1983), 141.
26. P. Laszlo and P. Pennetreau, Tetrahedron Lett., (1985), **26,** 2645.

27. P. Laszlo and J. Vandormael, Chem. Lett., (1988), 1843.
28. P. Laszlo and A. Mathy, Helv., Chim. Acta, (1987), **70,** 577.
29. J. H. Clark, A. P. Kybett, D. J. Macquarrie, S. J. Barlow and P. Landon, J. Chem. Soc., Chem. Commun., (1989), 1353.
30. A. Cornelis, P. Laszlo and P. Pennetreau, J. Org. Chem., (1983), **48,** 4771.
31. N. A. Noureldin and D. G. Lee, Tetrahedron Lett., (1981), **22,** 4889.
32. K. T. Liu and Y. C. Tong, J. Org. Chem., (1978), **43,** 2717.
33. E. Keinan and Y. Mazur, J. Org. Chem., (1978), **43,** 1020.
34. T. Nishiguchi and F. Asano, Tetrahedron Lett., (1988), **29,** 6265.
35. H. Ogawa, Y. Ichimura, T. Chihara, S. Teratani and K. Taya, Bull. Chem. Soc. Jpn., (1986), **59,** 2481.
36. M. Onaka, K. Sugita, M. Kawai and Y. Izumi, J. Org. Chem., (1989), **54,** 1116.
37. M. Onaka, M. Kawai and Y. Izumi, Bull. Chem. Soc. Jpn., (1986), **59,** 1761.
38. S. L. Regen, S. Quici and S. J. Liaw, J. Org. Chem. (1979), **44,** 2029.
39. S. L. Regen and C. Koteel, J. Am. Chem. Soc., (1977), **99,** 3837.
40. K. Smith, M. Butters and B. Nay, Synthesis, (1985), 1157.
41. J. San Filippo, Jr. and C. I. Chern, J. Org. Chem., (1977), **42,** 2182.
42. E. Santaniello and P. Ferraboschi, Nouv. J. Chim., (1980), **4,** 279.
43. K. Smith, M. Butters, W. E. Paget and B. Nay, Synthesis, (1985), 1155.
44. K. Sukata, Bull. Chem. Soc. Jpn., (1987), **60,** 1085.
45. E. Suzuki, K. Tohmori and Y. Ono, Chem. Lett., (1987), 2273.
46. E. C. Taylor, C. S. Chiang, A. McKillop and J. F. White, J. Am. Chem. Soc., (1976) **98,** 6750.
47. J. M. Melot, J. Texier-Boullet and A. Foucaud, Tetrahedron Lett., (1986), 27, 493.
48. D. Villemin, J. Chem. Soc., Chem. Commun., (1985), 870.
49. A. Corma, H. Garcia and J. Primo, J. Chem. Res. (S), (1988), 40.
50. M. J. Climent, A. Corma, H. Garcia and J. Primo, App. Catal., (1989), **51,** 113.
51. T. Nishiguchi and H. Taya, J. Am. Chem. Soc., (1989), **111,** 9103.
52. R. I. Hedge and M. A. Barteau, J. Catal., (1989), **120,** 387.
53. Y. Izumu, N. Natsume, H. Takamine, I. Tamaoki and K. Urabe, Bull. Chem. Soc. Jpn., (1989), **62,** 2159.
54. T. Nishiguchi and F. Asano, J. Org. Chem., (1989), **54,** 1531.
55. H. Konishi, K. Aritomi, T. Okano and J. Kiji, Bull. Chem. Soc. Jpn., (1989), **62,** 591.
56. J. Cabral, P. Laszlo and, L. Mahe, Tetrahedron Lett., (1989), **30,** 3969.
57. K. Smith, M. Butters and B. Nay, Tetrahedron Lett., (1988), **29,** 1319.
58. A. G. Mistry, K. Smith and M. R. Bye, Tetrahedron Lett., (1986), **27,** 1051.
59. K. Smith, Bull. Soc. Chim. France, (1989), 272.
60. P. Laszlo and H. Moison, Chem. Lett., (1989), 1031.
61. M. Hino and K. Arata, Chem. Lett., (1989), 971.

62. K. S. Kim, S. Chung, I. H. Cho and C. S. Hahn, Tetrahedron Lett., (1989), **30,** 2559.
63. E. Santaniello, F. Ponti and A. Manzocchi, Synthesis, (1978), 534.
64. T. Baba, K. Kameta, S. Nishiyama, S. Tsuruya and M. Masil, J. Chem. Soc., Chem. Commun., (1989), 1072.
65. M. S. Climent, J. M. Marinas and J. V. Sinisterra, React. Kinet. Catal. Lett., (1986), **32,** 177.
66. T. Nishiguchi and M. Bougauchi, J. Org. Chem., (1989), **54,** 3001.
67. T. Kimura, M. Fujita and T. Ando, J. Chem. Soc., Chem. Commun., (1990) 1213.
68. H. Liu, P. Wang and P. Sun, J. Fluorine Chem., (1989), **43,** 429.
69. T. Nishiguchi and C. Kamio, J. Chem. Soc. Perkin Trans., 1, (1989), 707.
70. T. Ando, S. Sumi, T. Kawate, J. Ichihara and T. Hanafusa, J. Chem. Soc., Chem. Commun., (1984), 439.
71. J. H. Clark and C. W. Jones, J. Chem. Soc., Chem. Commun., (1987), 1409.
72. L. M. Weinstock, J. M. Stevenson, S. A. Tomellini, S. H. Pan, T. Utne, R. B. Jobson and D. F. Reinhold, Tetrahedron Lett., (1986), **27,** 3845.
73. H. Alper, S. Ripley and T. L. Prince, J. Org. Chem., (1983), **48,** 250.
74. D. E. Bergbreiter and J. L. Lalonde, J. Org. Chem., (1987), **52,** 1601.
75. G. Suzukamo, M. Fukao and M. Minobe, Chem. Lett., (1987), 585.
76. M. Marczewski in Catalysis by Acids and Bases, eds. B. Imelik et al, Elsevier, (1985), 213.
77. T. J. Pinnavaia, R. Raythatha, J. G. S. Lee., L. J. Halloran and J. F. Hoffman, J. Am. Chem. Soc., (1979), **101,** 6891.
78. N. Davison and W. R. McWhinnie, Inorg. Chim. Acta, (1987), **131,** 9.
79. H. Sakurai, K. Urabe and Y. Izumi, J. Chem., Soc., Chem. Commun., (1988), 1519.
80. M. Onaka, R. Ohno, M. Kawai and Y. Izumi, Bull. Chem. Soc. Jpn., (1987), **60,** 2689.
81. J. A. Ballantine, W. Jones, J. H. Purnell, D. T. B. Tennakoon and J. M. Thomas, Chem. Lett., (1985), 763.
82. A. Cornélis, P-Y Herzé and P. Laszlo, Tetrahedron Lett., (1989), **23,** 5035.
83. P. Laszlo and J. Lucchetti, Tetrahedron Lett., (1984), **25,** 1567.
84. K. R. Kumar, B. M. Choudary, Z. Jamill and G. Thyagarajan, J. Chem. Soc., Chem. Commun., (1986), 130.
85. B. Chiche, A. Finiels, C. Gauthier, P. Geneste, J. Graille and D. Pioch, J. Mol. Catal., (1987), **42,** 229.
86. K. Urabe, J. Sakurai and Y. Izumi, J. Chem. Soc. Chem. Commun., (1986), 1074.
87. P. Laszlo and J. Lucchetti, Tetrahedron Lett., (1984), **25,** 2147.
88. P. Laszlo and J. Lucchetti, Tetrahedron Lett., (1984), **25,** 4387.
89. J. A. Ballantine, M. Davies, H. Purnell, M. Rayanakorn, J. M. Thomas and K. J. Williams, J. Chem. Soc., Chem. Commun., (1981), 8.

90. O. Horie and S. Okazaki, Chem. Lett. (1986), 1089.
91. M. Balogh, I. Hermecz, Z. Meszaros and P. Laszlo, Helv. Chim. Acta, (1984), **67,** 2270.
92. A. Cornélis and P. Laszlo, Synthesis, (1982), 162.
93. R. W. McCabe, J. M. Adams and K. Martin, J. Chem. Res. (S), (1985), 356.
94. D. T. B. Tennakoon, W. Jones, J. M. Thomas, J. H. Ballantine and J. H. Purnell, Solid State Ionics, (1987), **24,** 205.
95. A. Cornélis and P. Laszlo, Synthesis, (1980), 849.
96. A. Cornélis, N. Depaye, A Gerstams and P. Laszlo, Tetrahedron Lett., (1983), **24,** 3103.
97. J. M. Thomas, X. Kuan and J. Stachurski, J. Chem. Soc., Chem. Commun., (1988), 162.
98. A. Cornelis, A. Gerstmans, P. Laszlo, A. Mathy and I. Zieba, Catal. Lett., (1990). **6,** 103.
99. J. H. Clark, C. W. Jones, C. V. A. Duke and J. M. Miller, J. Chem. Soc., Chem. Commun., (1989), 81.
100. N. A. Noureldin and D. G. Lee, Tetrahedron Lett., (1981), **22,** 4889.
101. V. G. Kumar, T. S. Shoba and K. V. C. Rao, Tetrahedron Lett., (1985), **26,** 3281.
102. R. P. Singh, H. N. Subbarao and S. Dev, Tetrahedron, (1979), **35,** 1789.
103. S. L. Regen and C. Koteel, J. Am. Chem. Soc., (1977), **99,** 3837.
104. F. J. Kakis, M. Fetizon, N. Douchkine, M. Golfier, P. Mourgues and T. Prange, (1974), J. Org. Chem., **39,** 523.
105. P. Laszlo and E. Polla, (1984), Tetrahedron Lett. **25,** 3309.
106. J. H. Clark, A. P. Kybett, P. Landon, D. J. Macquarrie and K. Martin, (1989), J. Chem. Soc., Chem. Commun., 1355.
107. V. M. Thuy and P. Maitte, (1989), Bull. Soc. Chim. Belg., **98,** 877.
108. A. S. Radhakrishna, S. K. Suri, K. R. K. Prasad Rao, K. Sivaprakash and B. B. Singh, Synth. Commun., (1990), **20,** 345.
109. B. M. Choudary, V. L. K. Valli and A. Durga Prasad, J. Chem. Soc., Chem. Commun., (1990), 721.
110. L. Delaude and P. Laszlo, Catal. Lett., (1990), **5,** 35.
111. B. M. Choudary, A. Durgaprasad and V. L. K. Valli, Tetrahedron Lett., (1990), **31,** 5785.
112. P. H. J. Carlsen, M. Husbyn, J. E. Braenden and R. Eliason, Acta Chem. Scand., (1989), **43,** 485.
113. G. Bram, A. Loupy, M. Majdoub, E. Gutierrez and E. Ruiz-Hitzky, Tetrahedron, (1990), **46,** 5167.
114. B. M. Choudary, V. L. K. Valli and A. Durga Prasad, J. Chem. Soc., Chem. Commun., (1990), 1186.
115. H. Liu, H. Lu, P. Wang, P. Sun and F. Zhang, J. Chem. Soc. Perkin Trans. 1, (1990), 2624.

Index